D0687250

ENJOY OUR UNIVERSE

Albert Einstein holding the visible Universe, to whose current understanding he contributed so significantly. The visible Universe, © WMAP/NASA.

ENJOY OUR UNIVERSE

You Have No Other Choice

Alvaro De Rújula

Original illustrations by Alvaro De Rújula

OXFORD
UNIVERSITY PRESS

Great Clarendon Street, Oxford, OX2 6DP,
United Kingdom

Oxford University Press is a department of the University of Oxford.
It furthers the University's objective of excellence in research, scholarship,
and education by publishing worldwide. Oxford is a registered trade mark of
Oxford University Press in the UK and in certain other countries

First Edition published in 2018

Impression: 1

Published in the United States of America by Oxford University Press
198 Madison Avenue, New York, NY 10016, United States of America

British Library Cataloguing in Publication Data

Data available

Library of Congress Control Number: 2017962191

ISBN 978–0–19–881780–2

DOI: 10.1093/oso/9780198817802.001.0001

Printed and bound by
CPI Group (UK) Ltd, Croydon, CR0 4YY

Contents

Preface

Most people—and among them, most scientists—are familiar with the rules of some sports and the results of their recent games. This may be because they practice these sports themselves, or simply because sports are fun to follow. Most scientists practice their trade and, even if they do not, they consider that progress in science is fun to follow. And here ends the similarity between sports and science, or between "most scientists" and "most people."

A large fraction of non-scientists would not enthusiastically adhere to the opinion that understanding the Universe in which we happen to be, or simply to attempt to do it, is also great "fun." *I wouldn't understand anything* is not an uncommon reaction to any text on science. In my opinion, the main reason is not that science is boring and unfathomable but that, more often than not, it is not properly taught.

But even in a (good) kindergarten, kids may be taught the "scientific method"; doing experiments and drawing conclusions from their results. To be specific, fairly young children may be given a balance, a kettle, a kitchen utensil that measures volumes of liquid and various types of balls: a tennis-ball, a golf-ball, a table-tennis ball, a billiard-ball . . . The aim of the game is to figure out what it is that makes balls float or sink.

The results are surprising: Children find out fast that it is not the size that matters, but something also involving the weight. If they are not toddlers, given a small hint, or balls of the same size and different weight, they may even discover that the answer is in the relation of weight to volume (their ratio is the density). The game is more efficient—and teaches the benefits of collaboration—if the children are organized in teams. It also provides them with enjoyment, a useful way of thinking, and an approach to problems based on curiosity and constructive doubt.

The previous paragraph is not based on a *thought experiment*, a type of reasoning of which scientists are so very fond. It is the result of many "experiments" performed with—and by—real kids. There is a successful though insufficiently funded teaching program, pioneered by Leon Lederman (an American recipient of the Physics Nobel Prize) called "Hands On" (1). It employs precisely this trial and error—and try once again—teaching and learning methodology. It has been tested, in particular, with kids or youngsters from "problematic neighborhoods" in Chicago. The program has spread from the USA to other countries such as France, where it is called *La main à la pâte* (hands in the dough) (2). It was pioneered there by Georges Charpak, yet another Nobel Physics Laureate.

"Active learning" projects such as Hands On also exist for education up to secondary school or even university level. At the higher levels the problem may be that teachers have not been properly taught, or so says Carl Wieman—also a Nobel Physics laureate for a change—who has been very actively involved with this type of education (3). The point is to substitute medieval teaching techniques, essentially the endless repetition of texts to be memorized, with something more constructive.

This book is not intended for (very) young kids nor for physicists. It is intended for anyone—independently of the education (s)he suffered—who is interested in our basic current scientific understanding of the Universe. By "Universe" I mean everything observable from the largest object, the Universe itself, to the smallest ones, the **elementary particles** that "function" as if they had no smaller parts. This is one more of many books on the subject. Why write yet another one? Because the attempts to understand our Universe are indeed *fun* and I cannot resist the temptation of putting in writing—and attempting to partake of my own share of this fun.

Some potential readers may be allergic to math. Fear not. I shall use very little *algebra* (symbols to represent concepts, as in $E=m\,c^2$, or their relations, such as $\neq$, $\approx$, $\sim$, $\propto$, $>$, for not equal, approximately equal, roughly equal, proportional to, greater than) and *arithmetic:* the four rules, powers such as 10^3 (the cubic power of ten, aka ten to the third, aka one thousand). I may go as far as using an occasional *square root* or a *vector*, but not without warning.

Three extra advance notices. There will be a warning star or two in the title of the slightly more algebraic or unavoidably intricate chapters and sub-sections. They can be skipped according to the reader's preference. Some footnotes contain technical clarifications intended to pacify the most knowledgeable readers. At the end of the book there is a Glossary of Terms, since it would be difficult to memorize in passing all of the very many concepts and names introduced at one point and later reused, such as **fermion**, **boson**, and the names of the **powers of 10**.

References

[1] See, for instance, Looking Back: Innovative Programs of the Fermilab Education Office, Available online at http://ed.fnal.gov/office/marge/retro.html (accessed October 2017).

[2] Fondation de Coopération Scientifique pour L'Éducation à la Science. Website. Available at: www.fondation-lamap.org/

[3] Carl Wieman, *Scientific American*, August 2014, page 60.

Acknowledgments

I am indebted to Alicia Rivera, Fabio Truc, and five (!) anonymous OUP referees for their having carefully read the original manuscript and for many useful suggestions. I am particularly thankful to Elisa Muxella, who convinced me to write this book.

1

Physics as an Art Form

Should the Almighty have consulted me about Creation, I would have recommended something simpler.

ALPHONSO Xth of CASTILLE (1221–1284)
King of the Three Religions.

The royal quote was a commentary on the then accurate and pervading but very complex Ptolemaic view of the Solar System, in which planets revolved in complex orbits around the Earth, as opposed to simple ones around the Sun. This exceptional king, Alphonse the Wise, had good intuition: Things are indeed simpler!

For a scientist, "simplicity" is also synonymous with "beauty." The beauty, for instance, of the equations that describe the Universe. But the very first rule in writing a book on physics for readers who are not necessarily familiar with math is *not* to write any equations, with the possible exception of $E = m\,c^2$, which is so often misinterpreted.[1] Who can resist the temptation of breaking the rules? To camouflage my sinful manners, I break the rule by writing yet another equation, but in a figure, Figure 1.

The mystifying scribblings on Uncle Albert's lower blackboard in Figure 1 are the *Einstein's equations of* General Relativity. This rather simple formula encompasses a good fraction of "everything physicists know"—at a basic level.[2] It says that gravity is described by the properties of a gravitational field, embedded in the left-hand side of the equation; that is, before the "=" sign. The "source" of gravity (the right-hand side) is anything that has mass and/or energy and

[1] A point to which we shall dedicate the second part of Chapter 4.

[2] *Basic*, as opposed to *emergent*. The properties of a solid, for instance, "emerge" from the ones of the outer electrons of its constituent atoms. We know and understand an awful lot about emergent phenomena.

Enjoy Our Universe: You Have No Other Choice. Alvaro De Rújula.

DOI: 10.1093/oso/9780198817802.001.0001

Figure 1 Albert Einstein and his blackboards. On the lower one, the left-hand side describes gravity, the right-hand side its "source," discussed in detail in the second part of Chapter 7. The symbols c and G are the speed of light and Newton's constant, which determines the magnitude of the gravitational forces. μ and ν are "indices" that "run" from 0 to 3 for time and the three dimensions of space. The 8 and π are other rather well-known numbers. $g_{\mu\nu}$ describes the "geometry" of space-time, which more often than not is "curved," like the surface of a sphere. The symbols R are explicit functions of $g_{\mu\nu}$ and its space-time variations. For more about Λ, see the text.

momentum.[3] The equation describes or predicts—among other things and with incredible accuracy—Newton's proverbial falling apple, the behavior of clocks in GPS satellites, the orbits of the planets (including Mercury's and the peculiar "advance" of its perihelion[4]), the deflection of starlight by the Sun, the motions of stars and galaxies, the existence of black holes, the emission of gravitational waves by binary pulsars

[3] What these apparently obvious concepts actually mean is discussed in the beginning of Chapter 4.

[4] The point of its orbit closest to the Sun.

and black-hole mergers . . . all the way to the expanding Universe.[5] Who would not admit that there is a manner of beauty in this simplicity?

The upper blackboard in Figure 1 is an add-on by Einstein, who later thought that it might have been his biggest error: Yet, it may turn out to be one of his major contributions; not an easy contest to win. What Λ symbolizes is the Cosmological Constant. It is to be added to the other equation, and it may be interpreted as the "energy density of the vacuum." If Λ is not zero, the vacuum *is not void*, as we shall discuss in well deserved detail in the first section in Chapter 22. At the moment, the Cosmological Constant is the simplest explanation of the observed fact that the expansion of the Universe is currently accelerating. For a positive Λ, a "chunk" of vacuum in the Universe would gravitationally *repel* any other one, thus providing the acceleration. Is this not also "beautiful" or, to say the least, mesmerizing?

Incidentally, a deep question to which we have no serious answer is: why are the fundamental laws of Nature elegant and concise? Yet another Physics Nobel Prize winner, who unlike the ones in the Preface I shall not name, has a tentative answer: It is sufficient to look at our planet to conclude that, if Someone created the Universe, She did it at random. How could you otherwise explain, for instance, the political situation in many countries, including mine and perhaps yours as well? You may have noticed that my colleague is a feminist, to the point of assuming that the hypothetical creator is a Goddess. This Goddess happens to have good taste, in the sense that on weekends, when She is supposed to rest, She reads the previous week's physics literature. When She comes across something irresistibly beautiful—and I have given an example—She decides that it is The Truth; that is, an inescapable Law of Nature. Since the Goddess is almighty, the new Law of Nature becomes forever true . . . it even applies to the past when it was also unbreakable.

The statement that often the right answer—not only in science—is the simplest one is often referred to as *Ockham's Razor*, after the English Franciscan friar and philosopher (ca. 1287–1347). In many

[5] A knowledgable reader may quip that an extra hypothesis is required: the "Cosmological Principle," stating that, at all locations, the Universe has the same properties if averaged over sufficiently large volumes. Even more pedantically, I would answer that this is not an assumption, provided the Universe underwent a period "inflation" shortly after its birth, as discussed in Chapter 29.

Figure 2 William of Ockham with his razor and spoon.

cases—particularly in science—the tool to distinguish the simpler from the more complex hypotheses does not have to be as sharp as a razor, even a spoon would suffice, see Figure 2. A clear example is the heliocentric (Sun centered) view of the orbits of the planets versus the more complex geocentric one, where we are the navel of the Universe.

2

Science as a Sport

The quest for the laws of Nature is also a competitive "sport," very much akin to a race. One may think that getting it right is all that matters. For Nature being a player and also an infallible "referee," what else may matter? One problem is that, when the historical context is ripe, a specific discovery—be it theoretical or experimental—is often made almost simultaneously by more than one person or team. A classical case in fruitful mathematics is the invention of differential calculus (the use of minuscule steps to build a complete object, such as a planet's trajectory). Isaac Newton, Gottfried Wilhelm Leibniz, and their followers bitterly fought over who had it first and whether or not plagiarism was involved. Scientists competing for the same win and reaching it almost simultaneously lack a "photo-finish"—the infallible referee of running races—and the problem of priority is pervasive.

There is increasingly convincing evidence that the Vikings set foot in America as early as the tenth century. There is no question that the Amerindians were there much before that. And yet, the glory of "discovering" America goes to Columbus. Thus, the point is not being the first to discover something, but the last. I learned this from a colleague wise enough to admit that his most quoted work was only an improvement over previously known stuff. The moral is that most scientists are extremely sensitive to recognition, deserved or not; *mammy!, daddy!, look!*

The best ever series of scientific games must have been "Einstein vs. Nature." The first three games he won with apparent ease all in one year: 1905. In that *Annus Mirabilis* he discovered Relativity, understood *Brownian motion* and explained the *photoelectric effect.* Whether much of relativity was contained in previous work by Hendrik Lorentz, Henri Poincaré, Hermann Minkowski, and others actually did lead to a controversy, though none but Einstein wrote the best known result, $E = m\,c^2$. Brownian motion—the observed jitter of pollen grains in water—is due to the water molecules colliding with them. This conclusion

Enjoy Our Universe: You Have No Other Choice. Alvaro De Rújula.

DOI: 10.1093/oso/9780198817802.001.0001

Figure 3 Albert Einstein celebrating General Relativity.

definitely established the non-continuous "atomic" nature of matter. To explain in detail the photoelectric effect—how light impinging on metals extracts electrons from it—Einstein had to assume that light also occurs in "packets," the elementary particles that we now call photons. This earned Einstein a Nobel Prize, arguably not the most deserved one of the four he might have gotten.

The best of all Einstein-vs-Nature games was no doubt the one won by Einstein when he unveiled the enigma of gravity by figuring out the equations in Figure 1. The game started in 1907 with a disarmingly simple thought experiment: Einstein realized that someone inside a closed falling elevator would float in it and not know that (s)he is in a potentially fatal accelerating motion. From this he concluded that acceleration and gravity are "locally equivalent." The road from there to the equations of General Relativity is arduous even for someone studying them rather than attempting to work them out from scratch. It took Einstein up to 1915 to journey over it. We do not know how he felt upon reaching the road's endpoint, perhaps the way illustrated in Figure 3. For the rest of his life he struggled to find a "Unified Theory" of gravitation and electromagnetism, a task at which he failed.

3

The Bothersome Question of Units ★

The very many units used to measure the same reality, for instance distance, is esthetically interesting, but a real pain in practice. Not that one has to remember the exact relative magnitudes of miles, kilometers, nautical miles, verstas, royal feet, and what not, but, when traveling by car, it may prove useful to convert miles per gallon to liters for a hundred kilometers, or vice versa... and it is not so easy. Not to mention getting used to, say, what the upper fifties in Fahrenheit or six degrees Celsius *feel like* when they are not the units you grew up with. Perhaps, not so surprisingly, scientists choose 1—exactly one—as the value of as many fundamental quantities as they can.

In some respects, there has been progress concerning units. Nineteen countries of the European Union[6] and five other geographically small European countries[7] share a common currency: The Euro. This may or may not make economic sense, but it implies a gigantic simplification of banking, accounting, traveling, and so on. One may, for instance, cross most of (continental) Europe with a wallet, as opposed to a briefcase full of diverse banknotes.

Like currencies, the system of basic scientific units, such as those of length and time, has also evolved intelligently. The meter was originally defined as a fixed fraction (a tenth of a millionth) of the quadrant of the terrestrial meridian (the distance from the Pole to the Equator). Cross-country expeditions were once upon a time sent to "measure the meter," a delightful absurdity.

From 1889 to 1960, the meter was the length of a rod of platinum and iridium maintained somewhere in Paris at ice-melting temperature. The second used to be the fraction 1/(86,400) of the mean solar day, whatever that was. The speed of light, generally denoted with c, was a very large number of kilometers per second, with many digits and an

[6] Austria, Belgium, Cyprus, Estonia, Finland, France, Germany, Greece, Ireland, Italy, Latvia, Lithuania, Luxembourg, Malta, the Netherlands, Portugal, Slovakia, Slovenia, and Spain.

[7] Andorra, Cyprus, Monaco, San Marino, and the Vatican.

Enjoy Our Universe: You Have No Other Choice. Alvaro De Rújula.

DOI: 10.1093/oso/9780198817802.001.0001

inevitable uncertainty. Absurd, in particular, because the speed of light in vacuum is an unperturbable constant of Nature.

The *second* is now defined as a *fixed* number (9,192,631,770) of the periods of the light emitted by an atom of Cesium 133 in the "hyperfine" transition of its ground states,[8] which is yet another constant in Nature. Instead of looking for a ridiculous definition of the meter, the speed of light (in vacuum) is currently fixed to be *precisely* 299,792,458 meters per second. Combined with the current meaning of the second, this *defines* the *meter*, with considerable savings in platinum, iridium, craftsmanship, and travel.

Notice that, in the vacuum, in a nanosecond[9] light travels a distance of about one foot (29.9792458 cm, to be exact). If I were to define "my royal foot" as precisely that distance, the speed of light would be, in units of (my royal) feet per nanosecond, *precisely* $c = 1$. A fairly easy number to remember. Moreover, with $c = 1$, the units of time (nanoseconds) and distance (nanoseconds of light travel) are essentially the same, see Figure 4. Setting $c = 1$ is even better than measuring maritime distances

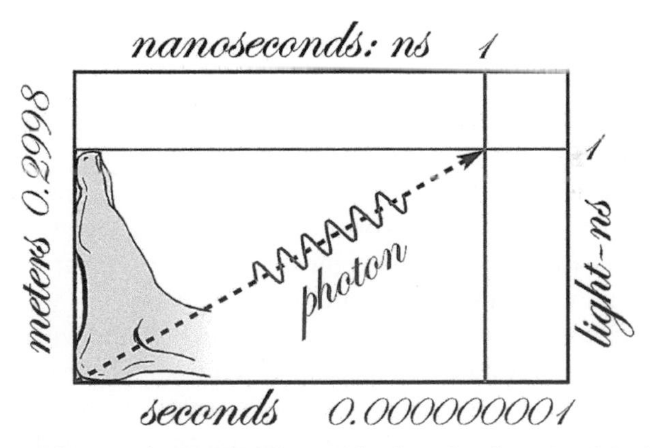

Figure 4 Light travels 29.9792458 cm (the length of *my Royal Foot*) in one nanosecond (ns) or, more simply, the distance of 1 light-ns in a ns. The horizontal and vertical scales are times and distances, respectively. Foot figure: courtesy of Pearson Scott Foresman.

[8] The two states correspond to the two relative orientations of the spin of the nucleus and that of the outermost electron. Spin is discussed in Chapter 11.

[9] One nanosecond (ns) is $10^{-9} = 1/10^9$ seconds. 10^9 is a (North) American billion; a British one is 1000 times more, a souvenir of the older Empire. We shall use the American convention here. A 1 followed by n zeroes is 10^n, its inverse is 10^{-n}.

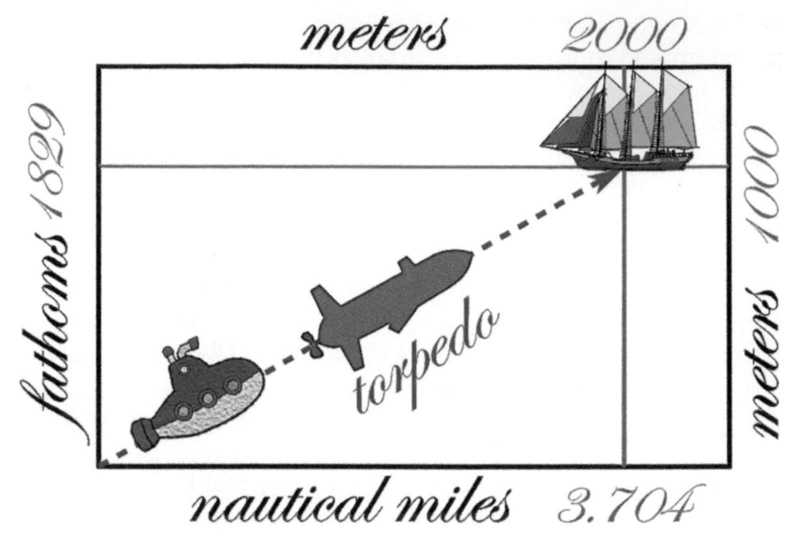

Figure 5 The measurement of distances in good old units (fathoms and nautical miles) or in meters. The horizontal and vertical scales are all for distances, in the various units. Ship: https://pixabay.com/en/ship-boat-sailing-travel-ocean-29244/ Submarine: https://pixabay.com/en/amarillo-blue-el-mar-mar-sea-sub-1300164/.

and depths in kilometers as opposed to nautical miles and fathoms; try to figure out, using the latter pair of units, the distance from your submarine to a ship at which you are aiming a (dummy!) torpedo. By the time you are done, the ship is gone. The units to be naturally used for torpedoes and ships are illustrated in Figure 5.

Physicists use **natural units**, fixing some fundamental quantities to the value "1." "Naturally," the speed of light is $c = 1$. For physicists, consequently, mass and energy are also measured in the same units: $E = m\,c^2$ simply becomes $E = m$. This does not mean, as we shall see anon, that Einstein discovered that the mass of an object changes with its energy, though he *did* say so introducing a misconception that, more than a century later, still lingers, even in textbooks.

4

The Scientific Method

> The truth is, the Science of Nature has been already too long made only a work of the Brain and the Fancy: It is now high time that it should return to the plainness and soundness of Observations on material and obvious things.
>
> Robert Hooke [1635–1703]

It is sometimes stated that a science is fundamental if the answers to the questions it poses *lie beyond* its established limits. That may be a fair description and—if one substitutes the word "fundamental" with "fascinating"—it may be the reason why some of the most successful popular scientific books are the ones that mainly discuss subjects that "lie beyond." In fundamental physics these subjects include the meaning of time, affordable travel to the past, dimensions beyond three (of space) and one (of time), universes other than ours, the very first instants of the Universe, the day "before," and so on.[10]

Except on rare occasions, I shall only deal with items that laboratory experiments or cosmological observations have shown to be correct. "Correct," as opposed to "true," in the sense that scientific progress is obtained as a succession of increasingly good approximations. When I deviate from this self-imposed limitation of sticking to the scientific method no doubt the reader will realize that I do it with tongue in cheek.[11]

[10] There is nothing wrong in discussing these subjects, except, in my opinion, doing it without a very clearcut distinction between facts, reasonable conjectures, and outright fantasies.

[11] An example: my interview of Einstein, whom I visit in his office in 1905. https://www.youtube.com/watch?v=iTWI-jV8HJY.

Enjoy Our Universe: You Have No Other Choice. Alvaro De Rújula.

DOI: 10.1093/oso/9780198817802.001.0001

Mass, energy, and momentum ★★

The discovery of the **Higgs boson** at **CERN**[12] announced on July 4th, 2012, had a totally unprecedented response from the media. Amongst the many reactions in the comments to the electronic versions of newspapers, there were endless discussions on the concept of "mass," triggered by the fact, which we shall discuss, that the Higgs **boson** has to do with the origin of the masses of (the other) elementary particles. This taught me that the rather simple concept of *mass* is often extremely well misunderstood, a misunderstanding to which Einstein himself significantly contributed. Consequently, I shall use this concept to illustrate the *scientific method.*

The method works like a simple, so far endless computer program:

1. Accept—perhaps temporarily—some concepts as fundamental.
2. Define them empirically (i.e., in practical terms).
3. Demand that the conceptual links between the concepts be consistent with observation.
4. Build up from there and, if this does not work, go back to (1) to start all over.

Let us, like mathematicians, be fastidiously precise, see Figure 6. And let us accept the concepts of time and space in the simple and practical sense that we can all agree about the measurement of time lapses and space distances with a given clock and a particular yard-stick. We shall say that an object covering the same straight-line distance in successive identical time intervals is in *uniform motion.* Now take a billiard table and some billiard balls. Clean and polish them to the point that the balls move uniformly on the table to the precision of our clock and yard-stick.

We are going to have to use an unavoidable ingredient of the scientific method: Math. Hopefully so little that one may choose not to pay particular attention to it. For a ball in uniform motion, or at rest, we can *define* a *velocity* as the ratio of a given distance and the time it takes the ball to cover it. On the table the velocity has two components, along its length (call it v_x) and width (v_y), which we can gather as a pair in a

[12] CERN is the European Laboratory for High-Energy Physics, located close to Geneva (Switzerland). An American competitor, Fermilab, is close to Geneva (Illinois). For more on these labs, see *CERN* and *High Energy Laboratories*, in the Glossary at the end of the book.

On their first visit to Venice, three people say:

The politician: "All gondoliers are women!"
The physicist: "This gondolier is a woman."
The mathematician: "All you can say is that the left-side of this gondolier ...
... seems to be that of a woman."

Figure 6 The diverse perceptions of different professionals.

vector,[13] $\vec{v} = (v_x, v_y)$. The next strenuous mathematical effort we shall make is to define the square of the velocity, v^2, as the sum of the squares in each direction: $v^2 = v_x^2 + v_y^2$. The length of the vector $\vec{v}$ is v, witness Pythagoras.[14]

Next, let us perform a series of experiments. We first take a collection of *identical* balls. We collide balls in motion with balls at rest or with other balls in motion and we make two fundamental discoveries:

- The *sum* of the velocities, $\vec{v}_1 + \vec{v}_2$ of a pair of colliding balls *before and after* the collision is the same, even if $\vec{v}_1$ and $\vec{v}_2$ change from before to after.[15] If unfamiliar with summing vectors, see Figure 7.
- The *sum* of the squared velocities of the colliding balls *before and after* the collision are the same.

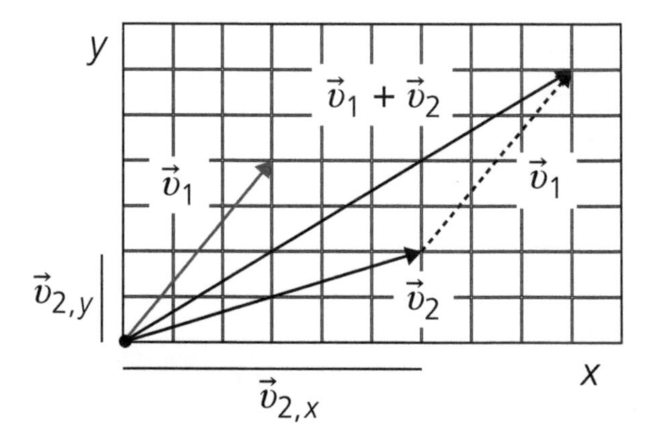

Figure 7 Two vectors, $\vec{v}_1$ and $\vec{v}_2$ and their sum. You may add their projections (or coordinates) along the horizontal axis and along the vertical one, to construct $\vec{v}_1 + \vec{v}_2 = (v_{1,x} + v_{2,x}, v_{1,y} + v_{2,y})$. But it is much simpler to "parallel transport" $\vec{v}_1$ to where it appears as dashed and then draw the line from the (black dot) origin to the tip of the dashed $\vec{v}_1$. The result is the vectors' sum. Constructed in this way $\vec{v}_1 + \vec{v}_2$ is obviously independent of the *coordinate system*, which may be rotated at will. Such a "system" is defined here by a choice of two *axes*, x and y, at right angles (90°; i.e., "orthogonal") to one another.

[13] A quantity having both a direction and a magnitude. From the latin for "carrier."

[14] Surreptitiously, I have introduced another mathematical concept; *the square root*. The square root of a number a is labeled $\sqrt{a}$. It is defined so that its square gives back the number: $(\sqrt{a})^2 = a$. For example, $\sqrt{4} = 2$. There is another solution: $\sqrt{4} = -2$, which, like $\sqrt{-4}$, we shall not need.

[15] The observation is that $\vec{v}_1^{in} + \vec{v}_2^{in}$ equals $\vec{v}_1^{out} + \vec{v}_2^{out}$.

Full of enthusiasm, we test our result with balls made of different materials but of the same size, so that the colliding balls still touch at the same height over the table. And we fail, the *conservation laws* that we had brilliantly discovered are false and we are to go back to square (1) of the scientific method.

Unshattered by failure, we try the next simplest possibility. We endow balls of different materials (labeled 1, 2,...) with a new property, a number. Call it m_1 for ball 1, m_2 for ball 2, and so on. And this time we are ready for two true discoveries; in collisions between balls 1 and 2 (or any other pair of balls) and for a *unique specific* choice of the ratio of m_1 to m_2:

- The sum of $m_1\,\vec{v}_1 + m_2\,\vec{v}_2$ *before and after* the collision are the same, even if $\vec{v}_1$ and $\vec{v}_2$ change from before to after.
- The same is true of the sum of $\frac{1}{2}\,m_1\,v_1^2$ and $\frac{1}{2}\,m_2\,v_2^2$: It stays put even if the collision changes the velocities.

Thanks to the scientific method, we have:

- Stumbled observationally upon two new fundamental concepts: *momentum*, $\vec{p} = m\,\vec{v}$, and **kinetic energy**, $E_k = \frac{1}{2}\,m\,v^2$.
- Discovered the "conservation" of the total momentum in a process (this time a collision): $\vec{p}_1^{\,in} + \vec{p}_2^{\,in} = \vec{p}_1^{\,out} + \vec{p}_2^{\,out}$.
- Discovered the "conservation" of kinetic energy, in the same sense.
- Empirically defined the concept of "**inertial**" mass,[16] m.
- Discovered in passing that the laws of particle dynamics are invariant under rotations, see the captions to Figures 7 and 8.

To give a parenthetical comment on what we have just discovered, let me recall a story from my time as a university student. We had a math professor who was much more evil than average, not an easy task at the location (Madrid) and time (not good, Franco was very much alive). One day, perhaps not spontaneously, he (the professor, not the "generalissimo") fell from the window of his third-floor office. Lying on the ground, half dead after the impact (which he survived) he kept on screaming "Thank God, it's only one half." After a while a daring student, immersed in the crowd gathered for the occasion, asked: "Professor, one half... one half of what?" "Idiot!... " said he (not uncharacteristically), "obviously... one half of em vee squared," that is $(1/2)\,m\,v^2$, with m and v the professor's mass and velocity.

[16] As opposed to "gravitational" mass, to be discussed as soon as we go back to gravity in the second part of Chapter 7 – to conclude that both masses are one and the same.

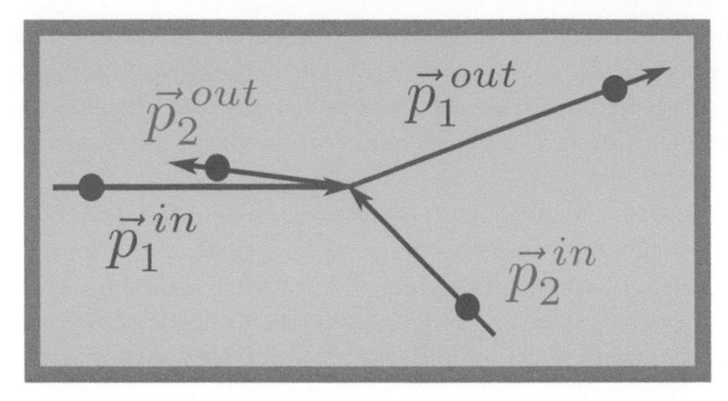

Figure 8 Colliding balls on a billiard table (ball 1 is red, ball 2 blue), *in* and *out* refer to before and after a collision. A billiards champion who trains by shooting his (red) ball to a moving (blue) one would tell you that, if the arrows' lengths reflect the ball's velocities, this collision is unreal. But the arrows reflect momenta (mass times velocity), not velocity. And, I now admit, ball 1 has twice the mass of ball 2. Given this, the illustrated collision "satisfies" momentum conservation: it is a possible one. Easy to check with the help of a pencil, Figure 7 and the liberty to use parallel transports.

Admittedly, I have cheated a little. I have not explained the factor of 1/2 in my professor's kinetic energy, which is to some extent a convention, and is related to the existence and customary definitions of other types of energy. One of them is the gravitational energy of the balls due to the Earth's gravitational pull, which I disposed of by (tacitly) assuming that the table was horizontal (and frictionless). There was a more serious swindle in that I gave no previous warning that the balls were "non-relativistic": Their velocities were negligible relative to the speed of light, within the precision of my instruments.

If the gravitational energy is much smaller than the kinetic energy and we repeated the collision experiments in mid-air (or better, in a vacuum) the momentum-conservation rules we find would be three-dimensional and not two-dimensional, like the surface our billiard table was. If gravity is not negligible, we would also discover the conservation laws, provided we took into account the gravitational energy differences at different heights.

The scientific method is universal, and unassailable. The experiments I have described could be performed by anybody, anywhere—thus the universality and the conclusions would be the same—thus the unassailability. *No other human creation or activity shares these strengths.*

The invention of the scientific method is generally attributed to Galileo Galilei, best known for his—presumably apocryphal—experiments on the dropping of objets from the leaning tower of Pisa: They fell at the same speed, independently of their weight. Even more memorable were his, very real, troubles with the Inquisition. A lesser known facet of Galileo is that he was a poet, and a funny one. His booklet *Contro il portar la toga* ("Against the donning of the gown") is written in iambic pentameters,[17] in spite of which it is absolutely hilarious. It describes how the university forced him to don his official gown on every outing, and how this was a problem when trying to visit, incognito, a brothel.

$E = m\,c^2$, right or wrong?★

For "**relativistic**" objects, moving at velocities v not much smaller than c, the conserved energy and momentum we have "discovered" are not the correct ones. If we played billiards with relativistic balls—as is routinely done at particle accelerators and colliders—we would discover that they are once more correct, provided we introduce a small modification: A velocity-dependent function, the **Lorentz factor**, $\gamma(v)$, is to be added here and there.

Let γ be $1/\sqrt{1 - v^2/c^2}$. Notice that if v is much smaller than c, $1 - v^2/c^2$ is approximately 1, and so is γ. Otherwise, the correct conserved momentum is no longer $\vec{p} = m\,\vec{v}$, but $\vec{p} = \gamma\, m\, \vec{v}$. The correct conserved energy is $E = \gamma\, m\, c^2$. This is *not* your ubiquitous $E = m\, c^2$. For some reason, in 1905, Einstein decided to redefine γm as "the mass," to conclude that it varies with velocity. But $E = m\, c^2$ (valid for particles at rest) is not an identity. Einstein misinterpreted himself.

We shall see that many particles are not stable, they *decay* into other particles. Atoms in a flame, for instance, are "excited" and they decay into "less excited" atomic states and the photons that we see as the shining of the flame. In such processes the initial energy (of the excited atom) equals the sum of the energies of its decay products. But the initial mass differs from the sum of the masses of the decay products. Thus, energy, which is conserved, and mass, which is not, are *not* equivalent. That is *the way it is*, observationally, and it ends the discussion of the presumed identity of mass and energy in an indisputable way.

[17] Lines containing five *iambs* (unstressed syllables followed by stressed ones).

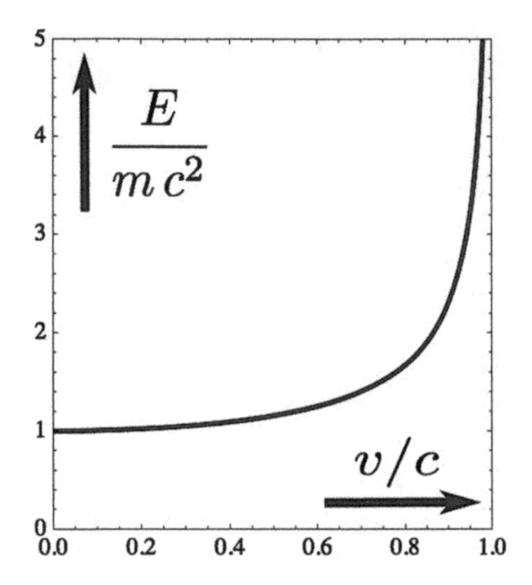

Figure 9 The energy, E, of a particle of mass m, divided by its rest energy, $m\,c^2$, as a function of its velocity v, divided by the speed of light, c.

In Figure 9 I have plotted $E = \gamma m c^2$ as a function of velocity (the velocity dependence is all in the factor γ). At $v = 0$, $E = m\,c^2$. At a small velocity, compared with c, γ is approximately $1 + (1/2)\,v^2/c^2$ [18] and E is approximately $m c^2$ (the "rest energy") plus E_k, the kinetic energy, $(1/2)\,m\,v^2$, which we "discovered" with the billiards experiment.

As v gets closer and closer to c, $\gamma = 1/\sqrt{1 - v^2/c^2}$ and $E = \gamma m\,c^2$ get closer and closer to infinity. It would take an infinite energy to accelerate a particle of non-vanishing mass to the speed of light. Thus the well-known limitation (do not travel faster than light!). The mass of photons—the particles of light that Einstein was the first to "invent"—is zero. Whatever their energy is, they travel at the speed of light.[19]

For v greater than c, γ is the square root of a negative number: it is not "real," but *imaginary*. So are the dreams of traveling faster than light. Or, even crazier, of Superman saving Lois Lane by *superluminally* flying to the past and "correcting" it.

[18] Trust me, if you do not know how to prove that this is a good approximation.

[19] A sufficiently "algebraic" reader may, with $c = 1$ to save on useless factors, check that for any object, *independent of its velocity relative to an observer*, $E^2 - |\vec{p}|^2 = m^2$ (which also holds for $m = 0$). In this sense mass is *relativistically invariant*, while energy and momentum are not. This reflects once again the *conceptual* difference between mass and energy.

5

The Three Relativities

To confuse the uninitiated, physicists distinguish between three theories of relativity: Einstein's "general" and "special" relativities and Galileo's relativity. The first is the theory of space, time, and gravity that we first encountered here when discussing Figure 1.

Galileo's relativity is a precursor of Einstein's special one. In a celebrated thought experiment, Galileo argued that two people playing ball in a closed enclosure on a ship would not know whether the ship is anchored or sailing at a constant velocity (he must also have tacitly assumed that the seas were not very rough). Thus, he concluded, the laws of motion must be independent of whether you are moving or not, relative to something else; the water in this case.

The boat is the ball player's reference system (or frame). For a vacationer lazily floating in the water, the water itself is "his" reference system (being lazy increases the probability that he is indeed a "he"). Such *unaccelerated* systems and the *observers* at rest in them are called *inertial*.

Special relativity is based on the assumption that all laws of Nature—including the ones governing light and electromagnetism—are the same for any *unaccelerated* observer. In particular, two such observers who are moving relative to each other would measure the same speed of light. You can emit a flash of light and then rush after it; you will still see it rushing ahead of you at a fixed velocity, c. Surprising, but correct.

A conclusion is that light is not moving in some universal fixed substance, or ether, a hypothetical entity defining an absolute space "absolutely at rest." Light is not a vibration of the ether. Unlike sound, which is a vibration of, for instance, air or your eardrums.

Unlike a constant velocity, an acceleration can be felt even with closed eyes.[20] General Relativity is a generalization of Special Relativity

[20] Everyone has had such feelings in a train station. One sees the neighboring train uniformly moving one way... or is it mine moving the other way? If this happened in empty space (trains but no station) there would be no way to decide. If *your* train accelerates, you do know.

Enjoy Our Universe: You Have No Other Choice. Alvaro De Rújula.

DOI: 10.1093/oso/9780198817802.001.0001

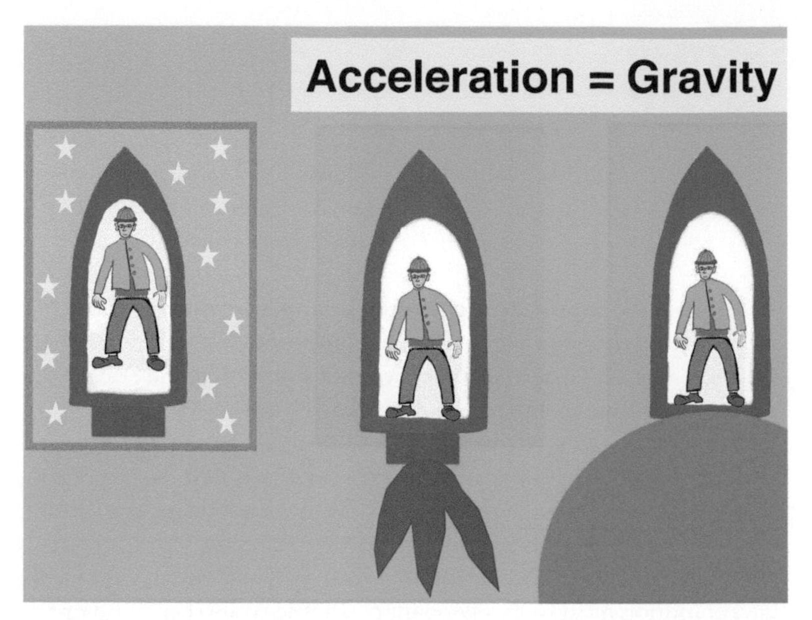

Figure 10 The guy in the rocket would be floating in it (if in a region with no significant gravitational forces), or if he and the rocket were in *free fall* in a gravitational field, as astronauts are in the Space Station (left). If he is in an accelerating rocket (middle) or at rest in a gravitational field (right) he cannot tell unless he cheats (by looking out of a window).

to include the physics of accelerations. Einstein postulated that, locally, acceleration and gravity are equivalent, as in Figure 10. Here, "locally" means in a region small enough for differences of gravitational forces at two different points to be unobservable. The tides are a counterexample.

The Twin Paradox *

I know exactly what time is... except if you ask me to explain it.

Augustine of Hippo [354–430]

(Perhaps significantly the patron saint of brewers)

Back to Special Relativity. Recall from the previous chapter that, in $c=1$ units, someone observing an object at rest would measure its energy as $E=m$, while, if the object is moving relative to the observer at a constant velocity v, the measurement yields $E=m\gamma$, with $\gamma=1/\sqrt{1-v^2}$.

Observations of time also behave in a velocity-dependent manner, though one must be much much more specific about how they are made.

Suppose one has placed two synchronized identical clocks in one's laboratory at a certain distance from each other and both at rest in the lab. We shall call the time they measure t_{rest}. A third clock, identical to the previous ones, is moving across the lab at a constant velocity, v. Let the time it measures be t_{mov}. When the moving clock passes by the first one at rest, their *stopwatch* buttons are activated, so that $t_{\text{rest}} = t_{\text{mov}} = 0$. When the moving clock passes by the second clock at rest, the *stop* buttons of the clocks are pressed. The prediction of Special Relativity is that at that moment $t_{\text{mov}} = t_{\text{rest}}/\gamma$. Since $\gamma > 1$, the moving clock *lags*!

This *time dilation* has been corroborated in many ways. Zillions of times, the "relativity" of time has been observationally checked, for instance, by observing the lifetimes of unstable particles. Looking at them as they move relative to the observer, they are seen to decay more and more slowly the faster they move, precisely by a factor of γ. Experiments corroborating this time dilation have also been done with precise clocks in airplanes. The GPS satellites would give the wrong time should they not know all this. There is no controversy here.

It is more intriguing to look at a clock moving in such a way that it flies by our first clock at rest (when they are set to $t_{\text{rest}} = t_{\text{mov}} = 0$) and then follows a path of constant velocity v, but changing direction, say traveling in a circle, bringing it back to the very first clock. If I have filled my lab with clocks at rest around the circle, I can measure how the moving clock ticks γ times more slowly in each bit of its trajectory, in comparison with the fixed clocks. Summing all the bits I predict the same result as before: The returning clock's time has been slowed down by a factor of γ. The prediction is correct. It leads to the *Twin Paradox* of Figure 11.

Why "paradox"? Reading the caption to Figure 11, no doubt you noticed that I seem to have, once again, cheated. I stated that the female twin ages faster than the male. Why did I not say that, from the point of view of the gentleman twin, it is the lady one who is traveling in a circle around him? From his point of view, is it not her who ought to be younger when they meet again? Two incompatible results!

The twin paradox implies that time travel is possible, though only to someone else's (e.g., my sister's) *future*. It may feel paradoxical, but it is a property of our Universe. The reason to state this with full confidence is simple: *Twin paradox experiments* prove it. Their detailed description will

Figure 11 Soon after we were born I flew away from my twin sister at a "relativistic" speed, v, such that $v^2/c^2 = 8/9$. By the time she reaches 60, I am only 20, since $\sqrt{1-8/9} = 1/3$. Had I covered the distance d by following a circle—as particles do in circular accelerators—we could meet and celebrate our 60th and 20th birthdays together. The twins: Max Pixel. FreeGreatPicture.com. Dürer's Mother by Albrecht Dürer (1471–1528). Michelangelo's David, a photo by Jörg Bittner Unna.

require using concepts that are only discussed in detail in later sections: "muons" and rings of magnets maintaining them in closed orbits. Thus, we shall only go back to the proverbial twins in Chapter 16, where we shall see that the twins' "story" is one of deep theory, beautiful thought experiments—as well as real ones—and a very thick plot.

6

A First, Fast Visit to the Universe

Imagine, for the sake of discussion, that you happen to arrive at our Universe, as in Figure 12, having somehow come from a completely different one where everything, including the laws governing Nature, is entirely different. Having the *unimaginable* technology required to indulge in this kind of tourism, there is no doubt that you are someone endowed with scientific curiosity. After looking around to find out "what there is" in our Universe and to understand "how things work,"

Figure 12 The Flammarion engraving, with someone trying to get out of our Universe and someone else coming from a different one. Camille Flammarion, *L'atmosphère Météreologie Populaire*, Paris 1888, p. 163.

Enjoy Our Universe: You Have No Other Choice. Alvaro De Rújula.

DOI: 10.1093/oso/9780198817802.001.0001

what will you conclude? Before we try to answer this question in more detail—and to the extent that, as of today, we have reached satisfactory answers—let us briefly see "who is who" and "what is what" in our Universe.

The Cosmic Background Radiation

Staring far into the sky, you (the alien traveler) would reckon that you have fallen in a sort of oven, in the sense that you are immersed in micro-waves. Very cold light waves, for they are identical to those in an oven whose inner walls are maintained at a temperature of about 2.73 K, with K standing for Kelvin.[21] The temperature of the humans you shall soon encounter is much higher: Some 309 K. The microwaves you detected are, like visible light, *electromagnetic radiation*, but they have a longer **wavelength**.

This microwave light uniformly permeates the Universe: Its waves consist of photons traveling in all directions at every point. It is called the **CBR (for Cosmic "Background" Radiation)**, and we detect it (with an adequate antenna) arriving from behind a lot of "foregrounds" (closer-by stuff); among other things more than a hundred thousand million (10^{11}) **galaxies**. One of them is "our" galaxy, the Milky Way. Once you manage to get rid of these foregrounds—as if you were using a wind-screen wiper—you see the place where the CBR that is now reaching you originated. It looks like the inside surface of a "celestial vault," shown in Figure 13, from which light took some 13.4 billion years to get to you.

The CBR—as we shall later discuss in extremely well deserved detail—was hotter in the past, as hot as the surface of the Sun. When referring to its current state at a temperature of ~2.73 K, it is often called the CMB (Cosmic Microwave Background). It is also dubbed MWBR (Micro-Wave Background Radiation), which might also stand for "More Will Be Revealed" (about the Universe).

Looking at a distant object is looking at the object as it appeared in the past, since light takes time to get from "there" to "here." When we see the Sun set, for instance, it actually went below the horizon some 8 minutes earlier: The Sun-Earth distance is approximately

[21] A Kelvin is a Celsius (or centigrade) degree above the *absolute zero* at −273.15 Celsius or −459.67 degrees Fahrenheit. At this temperature the thermal motions of atoms reach their minimum value.

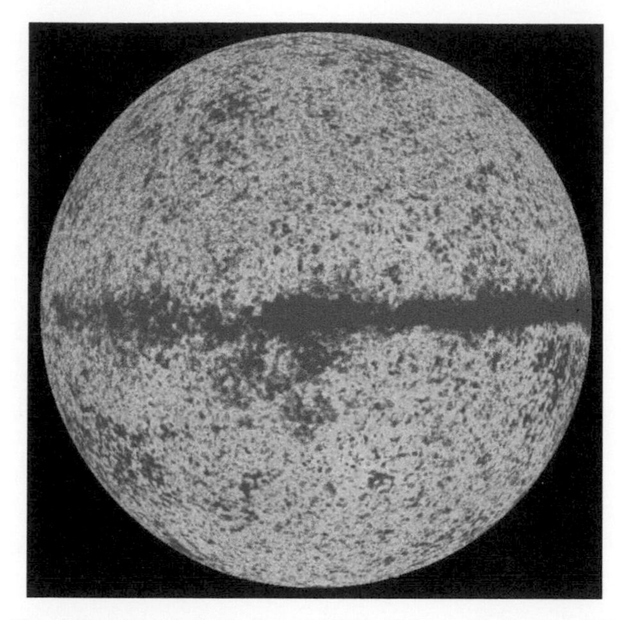

Figure 13 The Celestial Vault from which Cosmic Background Radiation reaches us. The red stripe is the emission from our own Galaxy, one of the "foregrounds." The direction toward the center of the Galaxy is in the middle, the rest corresponds to looking in all other directions. WMAP figure, courtesy of NASA.

8 light-minutes. Because the speed of light is so enormous, we do not "detect" it in everyday experience as we would with a sound's echo. In an example with sound, suppose that I am in Geneva, Switzerland. I kindly ask some of my many and very loud Italian friends to scream "ciao" at a precise time of the day. They are disposed as in the left-hand side of Figure 15. Since sound travels at about 1200 km per hour and they are spaced about 200 km from each other, I hear their "ciaos," fainter and fainter, at intervals of ten minutes ($200/1200 = 1/6$ and $1/6$ of an hour $= 10$ minutes).

But I happen to have many more Italian friends, and I now dispose them as in the right-hand side of Figure 14. At a given time, after their simultaneous screaming, I hear the ones marked in red all at once. If I add many of my other European friends to the exercise, the red guys in the figure span an entire circle. That circle is the sound equivalent of the CBR celestial vault in one fewer dimension of space (had I placed some of my friends above and below ground, the analogy would be complete).

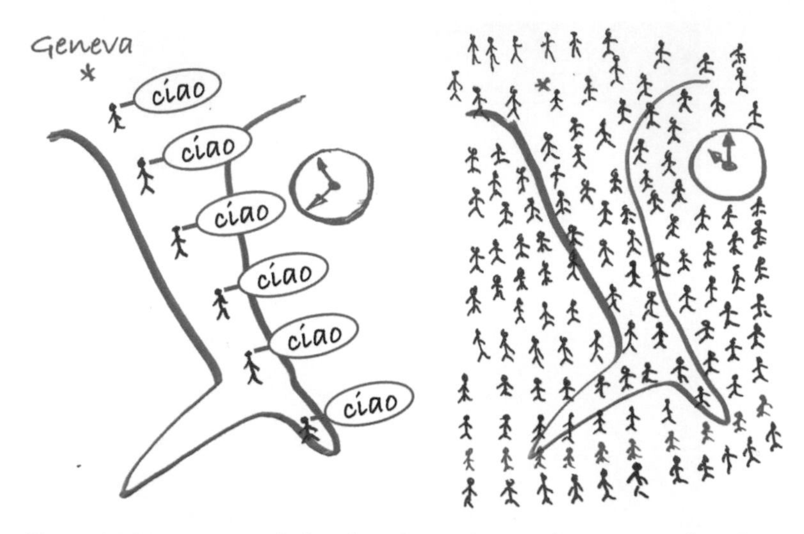

Figure 14 My numerous Italian friends simultaneously screaming "ciao."

Ten minutes later, the circle has a 200 km larger radius. The analogy is incomplete in that the celestial vault is not a sharp boundary, it is more akin to an orange's peel. And the Universe, unlike Italy, is observed to be expanding. Yet, in the future, the CBR will be reaching us from further away than it does now, from a place we have not yet seen.

The constituents of matter

Large or small, whatever else you see is lying closer to you than the CBR celestial vault. And everything there is in the Universe seems to be organized in a fairly simple way: At every scale there are more- or less-complex *particles* upon which *forces* act, as illustrated in Figure 15. In a first look at these things we shall find their names but not necessarily the precise meaning of these names. The meanings we shall hopefully learn little by little, like a child would, without apparent effort.

Moving from the largest toward the smaller objects, let us see what we find.[22] A galaxy-**cluster**—the biggest stabilized material "thing" we encounter—is made of galaxies, typically thousands of them. A very large galaxy, such as our neighboring Andromeda, shown in Figure 16,

[22] For the knowledgeable reader: when I say "see," I mean *directly* see; temporarily forgetting about "dark" matter.

Figure 15 All things made of "ordinary" matter. Typical sizes in meters are shown on the left-most column. Three of the four known fundamental forces are listed (the weak force does not play a role in stabilizing material objects). Nuclear forces binding protons and neutrons in nuclei are a consequence of the more fundamental Quantum Chromodynamic (QCD) interactions between quarks. Analogously, chemical forces between atoms are a peripheral manifestation of the Quantum Electrodynamic (QED) forces between electrons and nuclei.

Figure 16 Our neighboring M31, the Andromeda galaxy. Image courtesy of Adam Evans.

is made of a few hundred thousand million stars. Around a large fraction of these stars there are planets. *All* of the things mentioned in this paragraph are held together in their more or less stable configurations by the *force of gravity*, the very same force that prevents the reader from crashing into the ceiling when rising suddenly from a seat. Experts on these various large objects, in order of decreasing size (of the objects) are known as cosmologists, astrophysicists, and astronomers.

Living beings populate at least one planet. There is no scientifically defensible reason to assume that the circumstances are "just so" that planet Earth is a universal exception, the only one where there is life and even, to some extent, intelligent life.[23] The forces that briefly stabilize the groupings of "living particles," at least of the animal variety, are herd and sexual instincts. These are not "fundamental forces" and are, therefore, of no interest whatsoever to physicists. This time in order of increasing expertise, the experts on the "forces of life" are variously known as politicians, sociologists, biologists, and taxi-drivers.

[23] Define a scientific statement as one that can be *disproved* by observation. This ought to stop any disagreement on this point, at least until we explore all planets in the Universe or, less unlikely, some friendly E.T. comes to visit.

Particles at the next smaller-size level are *molecules* that, like water, H_2O, are composed of *atoms* (two of hydrogen and one of oxygen, in the most familiar chemical formula). Disrespectful of their name, atoms (Greek for *indivisible*) have parts. The helium atom, for instance, consists of two negatively charged *electrons* "bound" to a **nucleus** of doubly positive electric **charge**. The charge of an electron is -1, by definition.[24] Solids, liquids, molecules, and atoms are stabilized by the *electromagnetic forces* between their constituents. The immense variety of Nature on our scale of sizes—including all the complications of life, love, the rest of chemistry, mortgages, and tax-forms—ultimately stems from electromagnetic forces. The corresponding experts are known as condensed-matter physicists, atomic physicists, chemists, plumbers, and so on.

An *elementary particle*—by definition—is an object that has no parts, at least as far as we know at a certain moment in the development of science. As of today, electrons are elementary, like atoms once upon a time were thought to be. Electrons have been investigated down to distances about a billion times smaller than the size of an atom... and they stubbornly keep on acting as "point-like" particles.

The nuclei of atoms are not elementary, they are made of **protons** and **neutrons**, see Figure 17. Both these particles are, in turn, made up of two types of **quark**: The *up* quark and the *down* quark, to which we shall refer as u and d. These objects are not "up or down" in any geometrical sense, but the electric charge of an u is $+2/3$, "upward" of the charge of a d, which is $-1/3$. A proton, p, has charge $+1$, as it is made of two up quarks and a down quark $p = uud$. A neutron, $n = udd$, is neutral. Tacitly, as if it was evident, I have assumed that charges add up like numbers. Even worse, I have not (yet) reminded the reader of precisely what a charge *is*. Quarks, like electrons, are, as far as we know, elementary.

Once upon a time, scientists gave sophisticated appellations of Greek and/or mythological origin to the entities they hypothesized or discovered, such as electron, proton, atom, oxygen, helium, the names of the planets, and so on. The trend nowadays is to borrow words from common usage, with results that are often confusing and not always in the best taste. An example: The forces between quarks are called "glue" or "colored" forces, though these "**colors**" have nothing to do with the

[24] We have all learned by heart that opposite charges attract and may stick together.

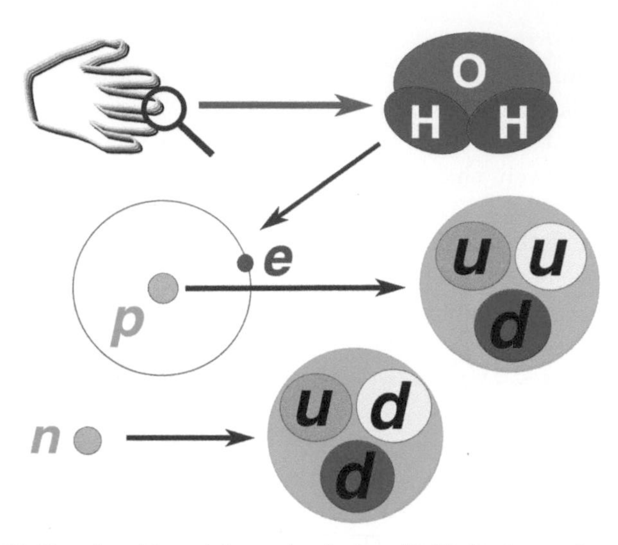

Figure 17 Your hand is mainly made of water (H_2O). Hydrogen is made of a proton and an electron. Quarks are the constituents of protons and neutrons. Electrons and quarks are elementary. The nucleus of the most common form (or *isotope*) of oxygen contains eight protons and eight neutrons. Isotopes of a chemical element have the same number of protons but different numbers of neutrons.

ones of light. Yet, there is a more elegant and a simpler alternative name: *chromodynamic forces* and strong interactions.[25]

All the ingredients required to make any material object, from a small atom to a large star, or even the most delicious apple pie, are up quarks, down quarks, and electrons, see Figure 17. To hold the ingredients together, all one needs are chromodynamic, electromagnetic, and (for a large object, such as a planet or a star) gravitational forces.

In the summary of all things made of ordinary matter[26] in Figure 15, there is a column on the *use* of the various objects. Quarks and the forces that bind them have no known specific practical application. That is why some enemies of science would say that the only use of quarks is to waste money ($) on the salaries of particle physicists. But knowledge is invaluable and the ability to understand how Nature works, or playing the violin, are amongst the few things distinguishing (some) humans from monkeys, see Figure 18.

[25] "Strong" used to refer to the no longer fundamental forces between the protons and neutrons of an atom's nucleus.

[26] There is also *dark* matter, which we shall discuss.

Figure 18 The symbol ≠ stands for "unequal." So are, to some degree, humans and monkeys.

The sizes of things

Elementary particles, such as a photon, an electron, or a quark have no size.[27] The smallest objects with size are the ones made of quarks, such as a proton. The largest observable object is the visible Universe (the Celestial Vault), whose radius is a bit more than 10^{28} cm. The geometric mean[28] of the largest and smallest objects we observe is approximately 10^8 cm, or 1000 km. That is the size of Spain, an "average" country.

It is not so easy to recall the relative sizes of all things material. A bit easier on a scale in which, instead of saying that, for example, the size of an atom is 10^{-8} cm, one simply writes "−8."[29] The sizes in centimeters of various objects are listed in Figure 19.

In a sense we live in a small Universe. The surprisingly large objects are the galaxies. Compared to them, the Universe is only "a bit" larger. This is illustrated in Figure 20, wherein the chosen "meter stick" is our

[27] More pedantically, the interactions between the fields that describe their behavior, discussed in Chapter 13, are "local": They occur at a fixed point in space-time.

[28] The geometric mean (or geometric average) of two numbers x and y is the square root of their product $\sqrt{xy}$.

[29] Such a scale is called "logarithmic" in base 10 for, by definition, $\log_{10}(10^n) = n$ (meaning that n is the base-10 logarithm of 10 to the power n). "Taking a logarithm" undoes the operation of "raising" to a power.

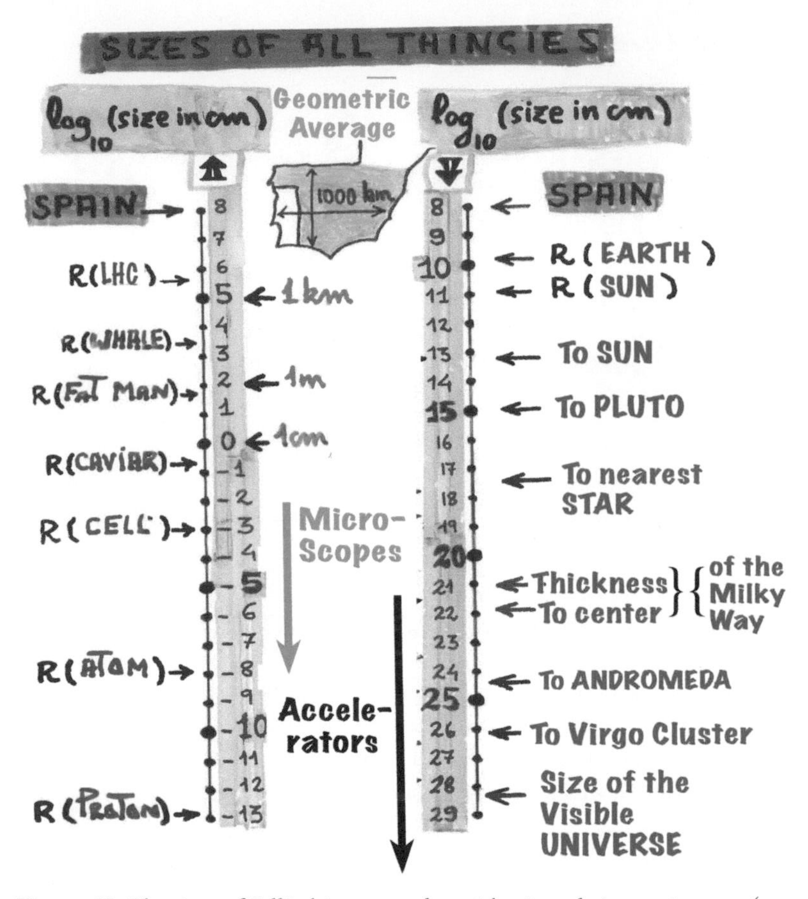

Figure 19 The sizes of "all" things, on a logarithmic scale in centimeters (see footnote 29). The LHC (the Large Hadron Collider at CERN, near Geneva, Switzerland) has a radius of about 7 km. A caviar egg, a few millimeters. Spain has a geometric (or logarithmic) average size, in the ensemble of all sizeable things.

Galaxy, whose full size is represented as that of a 5 Swiss Franc coin.[30] This choice was tempting, the Swiss Franc (SF) being a fairly stable gauge, like the good old platinum/iridium "official" meter. On that scale, the larger entities in the figure are an SF-throw away.

[30] Whose diameter is 3.145 cm, to be unnecessarily precise.

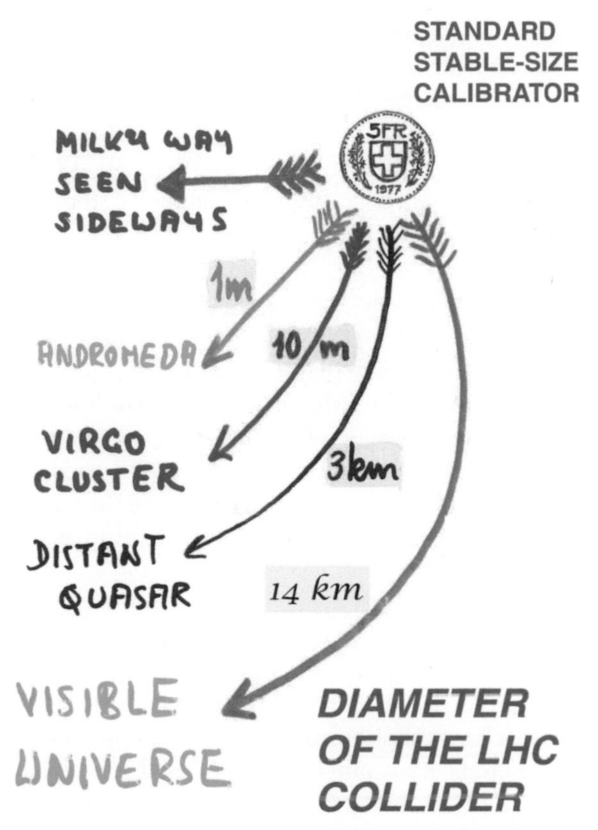

Figure 20 Large distances, compared to the full size of our Galaxy, represented by a 5 Swiss-Franc coin. A quasar is a gigantic black hole, visible from afar given the intense luminosity of the jets it spits out as it swallows the central areas of its "host" galaxy. *Virgo* is the cluster of galaxies to which we belong. The ratio of the sizes of the Galaxy and the visible Universe equals that of the coin and the LHC.

The first additional character: The neutrino

The ensemble $\{u, d, e\}$ of the two quarks and the electron is the list of ingredients needed to make anything directly visible, from a hydrogen atom to a cluster of galaxies, but it is not complete and it does not address the question of how "things work."

Natural radioactivity is a process involving several odd features of Nature. An example of a radioactive process is the *decay*, $n \to p e \nu$, of a neutron into a proton, an electron, and a neutrino, ν. More rigorously, a d quark in a neutron decays into an u quark ($d \rightarrow u e \nu$), which ends up as a constituent of the final-state proton. The process, called a "weak decay" is due to a force we have not listed: *The weak force*. And it is an example of something (the neutron) that "breaks into pieces" ($p e \nu$); peculiar "pieces" that are born—created—in the process, not honest-to-goodness pre-existing pieces of a neutron that would literally "break" into its pieces.

Neutrinos have no electrical charge. Unlike electrons, they do not bind to nuclei to form atoms. Although they are consequently "useless" as constituents of matter, they play a crucial role in our everyday life. Indeed, the first of the main series of reactions that "fuel" the Sun is essentially the same as "backward" neutron decay: Two protons and an electron "burn" into the nucleus of deuterium (pn) and a neutrino, $epp \rightarrow (pn)\, \nu$. Thus the instruction manual to operate the Sun *requires* the existence of neutrinos. And so do all forms of life ultimately powered by solar light.

Neutrinos "suffer" only two types of interactions: Gravitational and weak. As a consequence, neutrinos, in common circumstances, are extraordinarily penetrating, almost ghost-like. Compare, for instance, a beam of neutrinos with an energy of 1 GeV, roughly the rest energy of a proton,[31] with a beam of photons of the same energy, as in Figure 21. Incidentally, photons have different names at different energies, at 1 GeV they would be *gamma rays*, while microwaves and radio waves, infrared, visible, and ultraviolet light—as well as X-rays—are more familiar. It would take a mere 3 cm of water to absorb 50% of this gamma-ray beam. To absorb the neutrinos would require quite a swimming pool, as explained in the figure and its caption.

For decades, beams of evasive neutrinos have been made at particle accelerators, machines that we shall later discuss. These beams are directed to detectors, wherein a tiny fraction of neutrinos interact, creating particles that are much easier to observe. Much of what we know about neutrinos is extracted, indirectly, from the analysis of these interactions. For years, a neutrino beam produced at CERN, after having crossed two or three detectors, traversed (twice) the French Jura sierras, as in Figure 22. The local farmers complain about everything

[31] An *electron volt* (eV) is the energy acquired by an electron when accelerated from rest up to the positive plate in an electrical potential difference of one volt. A keV, MeV, GeV, TeV are 10^3, 10^6, 10^9, and 10^{12} eV, respectively.

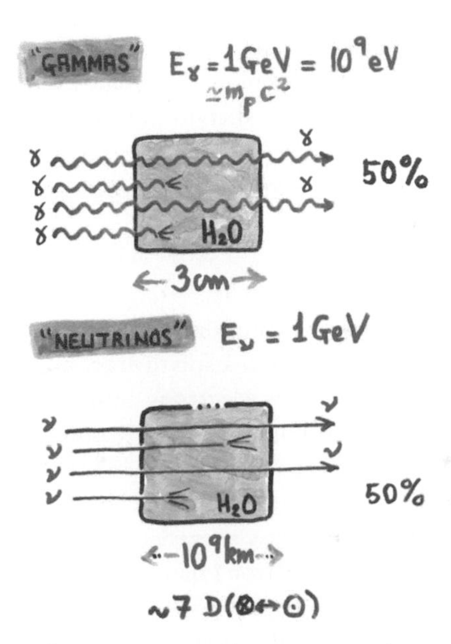

Figure 21 A beam of gamma rays with an energy of 1 GeV would be 50% absorbed in 3 cm of water. For neutrinos, that would take a water pool as long as seven times the distance, *D*, from the Earth (⊗) to the Sun (⊙).

Figure 22 A ghost-like CERN neutrino beam: It crosses a neutrino detector and gets lost in the skies after going twice through the Jura mountains.

imaginable, except the entirely negligible effects of neutrinos. Their grasslands, cows and goats continue to produce excellent milk, the basis of unparalleled cheese varieties such as Comté.

Doubling the cast of players, an interlude on antimatter

Quantum mechanics and relativity—subjects to which we shall come back—were the two great revolutions of physics in the early twentieth century. In 1928 Paul Adrien Maurice Dirac, an extremely taciturn Englishman, investigated how electrons could obey the rules of *both* of the then new theories. For the wrong reasons he stumbled on the right answer: It was possible only if the electron had an **antimatter** counterpart with the opposite electrical charge. Dirac supposed that that was the proton, but it was soon realized that the **positron** (the antimatter sibling of the electron) had to have the same mass as the electron. But protons are roughly 1836 times more massive than electrons. The positron (e^+) was discovered by Carl Anderson in 1932 and, indeed, it has the same mass as the electron (to a current precision of better than 1 part in 10^8).

For the same reason that the existence of the electron (e or e^-) requires that of the positron ($\bar{e}$ or e^+), there must be antiquarks ($\bar{u}$ and $\bar{d}$), and, consequently, *antimatter:* Antiprotons, antinuclei, anti-atoms...Neutrinos do not have electrical charges, but they have "weak" charges. That also implies that there must be two distinct types[32] of neutrinos: ν and $\bar{\nu}$.

Photons have no charge and are neither matter nor antimatter. When a positron and an electron meet, they *annihilate* into two or three photons, as in the *Feynman diagrams* of Figure 23. These days, at places such as CERN, anti-atoms are also routinely produced. In particular, anti-hydrogen: A bound system consisting of an antiproton and a positron. In contact with ordinary matter (for instance, hydrogen), antihydrogen also annihilates. Energy is conserved in the process and twice the rest energy of hydrogen is transferred to the resulting annihilation products. This releases about a billion times the kinetic energy of the final products of a chemical reaction, such as the burning of one atom of carbon and two of oxygen to make CO_2.

[32] We do not yet know yet whether the matter-antimatter relationship is precisely the same for neutrinos as it is for electrons and quarks, as discussed in Chapter 31.

Figure 23 Richard Feynman and some of his drums and diagrams. In the gray area, the annihilation of an electron and a positron into two or three photons (the wavy lines). The particle (vertically) exchanged is an e^+ or an e^-, depending on how the overall process is ordered in time.

In the process of $e^+ e^-$ annihilation into photons the initial charges add up to zero and so do the final ones, the photons being neutral. This is an example of *charge conservation*, which is an inviolate law of Nature.

There being no *antimatter mines,* antimatter is not a very useful fuel, it takes—in ideal conditions of 100% efficiency—as much energy to make it as it would be delivered when "using" it. Moreover, for obvious reasons, antimatter is extraordinarily difficult to handle. Too bad for CERN-inspired science fiction.

How elementary is an elementary particle?

The opinion is often stated that elementary particles are so difficult to understand because the subject is so complex, worse than professional wine tasting. But, unlike an excellent wine, elementary particles are difficult to grasp... because they are so astonishingly *simple*. An electron,

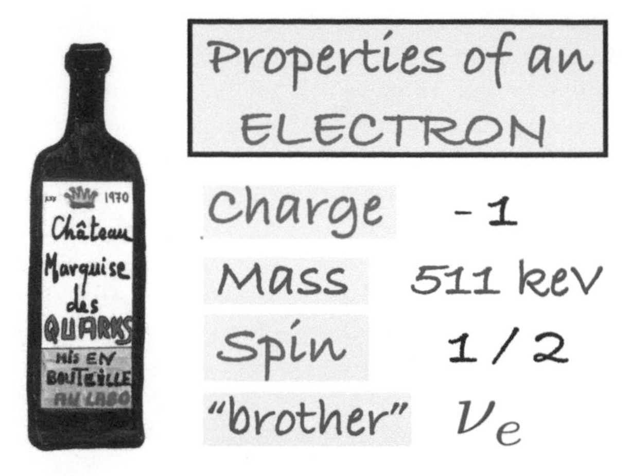

Figure 24 The four basic properties of an electron: Charge, mass, spin, and a "brother," ν_e, the neutrino "of the electron" (all concepts to be discussed in the text). Fully describing a good wine might take a bit longer.

Figure 25 Two similar chaps differ in their mustaches.

for instance, has only *four basic* properties[33], listed in Figure 24 and whose meanings are explained earlier or later in the text.

Like Thomson and Thompson (Dupont and Dupond, in Hergé's original version) the characters in Figure 25 are not even brothers. Neither are they identical, their mustaches are different. Contrariwise, two elementary particles with the same name, for instance two electrons, are fully identical and absolutely indistinguishable in principle and in practice. There are no bald or hairy, young or old electrons. Perhaps this is their most unfamiliar property, relative to the "macroscopic" objects we encounter every day.

[33] Another property, the *magnetic dipole moment*, is calculable from the basic ones. It is the amount by which an electron acts as a tiny magnet, as discussed in Chapter 13.

7

More on the Fundamental Forces and Their "Carriers"

We have quoted the four known fundamental *forces* or *interactions:* electromagnetism, gravity, chromodynamics, and the **weak interactions**. What *are* they?

Electromagnetism

The hydrogen atom, H, which is the simplest, consists of a proton and an electron bound to each other: $H = (e, p)$. What links them is an attraction between opposite electric charges: +1 and −1 for the proton and electron, respectively. This attraction is not a mysterious action at a distance—it has a go-between that transmits it from one particle to the other, as shown in Figure 26. This binding go-between is yet another disguise of an elementary particle: the *photon*, γ.

The *electric charge* is a property that allows an object to emit and absorb photons. The simplest *electromagnetic interaction* is the emission or absorption of photons by a charged particle. The *interchange of photons* between charged particles gives rise to the *electromagnetic force,* of which the photon is the *carrier*. There are plenty of related concepts to describe electromagnetism, the simplest non-trivial ones being charged particles and the specific way (proportional to the charge) in which they "couple to" (emit or absorb) photons.[34] Specify this specific way to a competent physicist and (s)he can derive all properties of atoms, matter, cellphones, and what not. Finally, the "size" of a charge describes the magnitude of its **coupling** to a photon. That is what the electric charge *is.*

[34] For the purist, I should have said that photons are described by a *quantum field,* the *source* of which is an *electromagnetic current,* whose specification may involve entities other than the charge, such as a dipole electric or magnetic moment.

Enjoy Our Universe: You Have No Other Choice. Alvaro De Rújula.

DOI: 10.1093/oso/9780198817802.001.0001

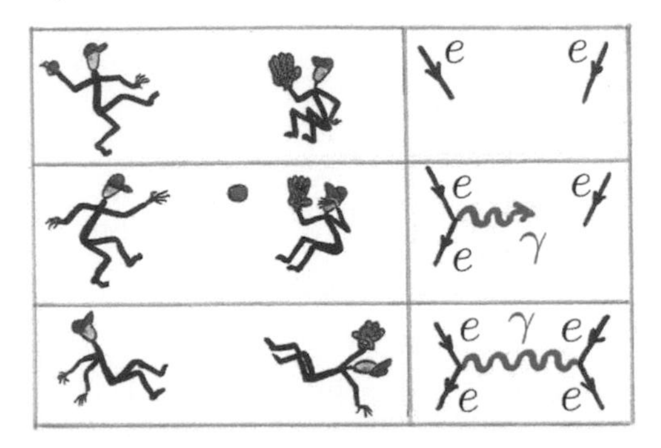

Figure 26 Interaction between two ball players (or electrons, e, the straight lines) by the exchance of a ball (or a photon, γ, the wavy line).

There is a common confusion, which I have not yet tried to dissipate, between the relative and actual magnitudes of a charge. When saying that the charge of an electron is -1, what one means is that it is equal and opposite to the charge of a proton, $+1$. Saying that the charge of an up quark is 2/3 of one means is that it is 2/3 of the charge of a proton. None of this refers to the actual value of the "size" of a charge. For electrons and protons (and up to the sign difference) the size of their charge, in natural units, is $e \sim 0.3$. You never see it written this way; physicists prefer to quote what they actually measure, the quantity $\alpha = e^2/(4\pi)$. Its value is $\alpha \simeq 1/137$, you can check that it implies $e \simeq 0.30286$. To open a physicist's lock, try 137. It is the number we remember best.

Hydrogen also exists briefly in "excited" levels (H^*) of energy higher than its so-called "ground state." An excited state can decay into a less excited one (or into the "ground" state, H) by the emission of a photon, as shown in Figure 27. An excited state has a larger mass than a less-excited one. In this process, the emitted photon is a particle that can travel merrily, just like any other stable particle. Thus the photon can act both as a force and as a particle; it is a *dual* entity. This "saving" of basic concepts, the understanding of two things "for the price of one," is called a *unification*. A unification is one of the most radical forms of scientific progress, an Ockham's delight.

A fascinating feature of quantum mechanics—thus, of the Universe we live in—is that photons, like any other particles, can behave as individual objects (particles, strictly speaking) or as *waves*. The defining

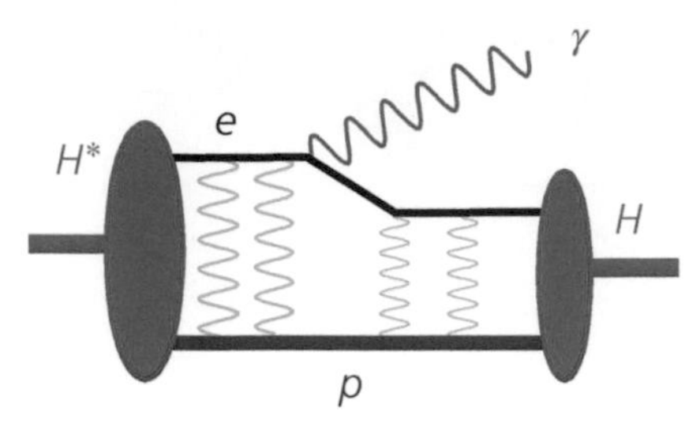

Figure 27 Decay of an excited state of the hydrogen atom, H^*, into its ground state H and a photon γ. Photon interchange is also responsible for tying the oppositely charged electron, *e*, and proton, *p*, in the *bound states* H^* and H; Two *energy levels* of hydrogen.

characteristic of waves is that they can add or subtract. Two stones dropped in a pond will produce circularly expanding water waves. When these waves meet, they generate a characteristic *interference* pattern, whose crests and troughs are (if the stones are identical) twice as large as those of the individual waves.

The *wave-particle duality* is yet another fundamental unification. Even an individual photon can behave—simultaneously—as a particle (which appears to follow a line trajectory) and a wave (which does not). This is so peculiar that we shall have to come back to it; seeing is believing. Incidentally, electrons, or any other objects, also have a wave and a particle aspect. This was first fully understood regarding photons by Einstein, and this is the reason why he got a Nobel Prize.

In all of the previous, the ending "magnetic" follows the prefix *electro-* because electricity and magnetism are also aspects of one basic thing: The Magnetism is an effect, also transmitted by photons, of electric charges in motion. Yet another example of unification: Nature is indeed miserable in her choice of basic entities.

The theory of the interactions between photons and electrons (and other charged particles) is currently called *QED*, for "Quantum Electro-Dynamics." To judge from the precision of its very many correct predictions, it is the most successful physical theory ever. The predicted value of the *magnetic moment* of the electron, for instance, agrees with experiment to better than 1 part in 10^{12}, as elaborated on in Chapter 13.

The remaining fundamental forces are analogous to electromagnetism. That is, they are also interchanges of particles.

Gravity and the "inertial" and "gravitational" masses

The most familiar force—and often the most annoying—is gravity. Even while lying down in a dark room, you know which way is "up." The room is permeated by something that you cannot smell, hear, taste, touch, or see, but of which you are definitely aware. It is the Earth's gravitational force. It holds you down in your bed, and it keeps the Moon in her orbit. You could also stay "up there" in orbit, should you be moving like the Moon or an astronaut at an adequate "orbital velocity," tangential to your trajectory.

For non-relativistic objects, whose relative velocity is much smaller than that of light, like the Moon and the Earth—or the latter and your self, should you be just walking around—the gravitational "charge" (analogous to the electric one) is their *mass*. The particle that *carries* the force of gravity is called the **graviton**.

For entities with relative velocities *not* negligible relative to *c*, the role of the gravitational "charge" or "source" is played by the *energy-momentum tensor*, $T_{\mu\nu}$ in Figure 1. For a non-rotating particle or star, this object is a simple and explicit function of its energy, its momentum and its location. With the object brought to rest, $T_{\mu\nu}$ becomes a function of its location and its *inertial* mass, which we "discovered" in the first part of Chapter 4. The concept of *gravitational mass* does not have an independent meaning, *gravitational and inertial masses are one and the same*—at least until Einstein's theory of gravitation is proved wrong. It makes some sense to say "heavier" when one means "more massive."

Gravity is the force we understand least, even if it was the first we understood something about. "We," here, means Isaac Newton, the reader and I, see Figure 28. From the experimental point of view, this is due to the fact that gravity—unless it is exerted by a large body like the Earth—is a very weak force, so that the observation of the emission or absorption of a *single* graviton is for the moment technically out of the question. We have not yet seen what **spin** means, let me mystifyingly just say that the "large" spin of the graviton (2) makes the theoretical construction of a consistent quantum theory of gravity much harder than it does for electromagnetism (the spin of the photon is merely 1).

While gravitons have not been individually detected, gravitational waves have. As we shall discuss in detail in Chapters 18 and 19 there are pairs of **neutron stars** rapidly orbiting about each other and *black hole*

Figure 28 Newton's law of the equal but oppositely directed forces of action and reaction [top]. Newton's apocryphal apple leading him to discover the relation between force, mass, and acceleration ($F = M a$), [center; expletive deleted], as well as the expression for a gravitational force between two objects, in terms of their masses and the distance between their centers [bottom]. As in Figure 1, G is "Newton's constant."

pairs merging into a single one. Much as an electron oscillating in a radio antenna emits electromagnetic waves, these objects emit gravitational waves, precisely as predicted by the equations in Figure 1.

The weak force

A neutron in a nucleus can *decay* into a proton, an electron and a neutrino, as in Figure 29(a). Here I employ the usual convention that this is an *anti*-neutrino, and the corresponding barred notation, $\bar{\nu}$. If looked at "more closely" the process is the decay of a down quark, $d \to u\, e\, \bar{\nu}$, see Figure 29(b). Analyzed even more closely, as in Figure 29(c), the process is mediated by a charged *intermediate boson*, called a W^-. This "boson" and its antiparticle, W^+, are carriers of the weak force; so called because under ordinary circumstances it is weaker than electromagnetism.

We have seen that a process mediated by these *weak intermediate bosons* is responsible for the operation of the Sun, thus the growth of grapevines,

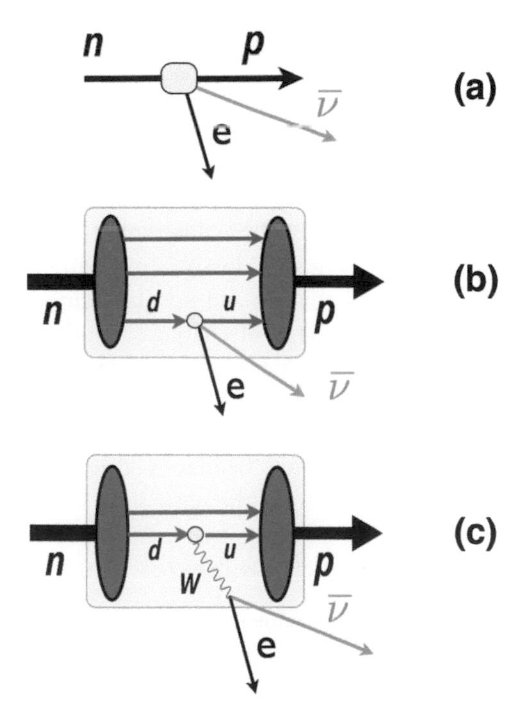

Figure 29 Neutron decay seen with increasing degrees of resolution.

and, ultimately, the existence of wine. Yet, *W*, as in these bosons, stands for "weak" not for "wino." This is not an entirely stupid joke of mine, for there is a hypothetical particle called the *Wino* that mercifully—in a linguistic sense—has not (yet) been discovered.[35]

There exists yet another weak intermediate boson, this time electrically neutral, the Z or Z^0. Being a neutral elementary particle akin to the photon, it has no antiparticle. The weak bosons, their masses, and the way they interact with each other, and with quarks, electrons, and neutrinos, were all correctly predicted well before they were discovered. All of this—in the context of the unification of electromagnetism and the weak interactions—is soon to be discussed.

Chromodynamics, the strong interaction

The fourth known fundamental force of Nature is the *strong force*, responsible for the binding of the quarks that constitute protons and neutrons. It has a plethora of different names ranging from the sophomoric *glue*, to the sophisticated *Quantum ChromoDynamics (QCD)*, and including the *strong interaction*.

QCD is analogous to QED. The carriers of the QCD interactions are particles, called gluons, similar to photons. The role played in QED by the electric charge is performed in QCD by something bemusedly called *color*. The "color charge," color for short, is also absolutely conserved. Each quark comes in three different colors. There are, for instance, three u quarks, whose colors we may arbitrarily call red, blue, and yellow. Electrical charges add up like numbers. A neutron, for instance, is made an u quark of charge 2/3 and two d quarks of charge −1/3. The charge of the neutron, $Q[n]$ is zero, as illustrated in Figure 30.

The quoted figure sounds less blatantly obvious if one realizes that there is another way to "get zero," besides making a step forward and two half-steps backward, all along the same line. In a plane, one can also go back to the starting point (get zero displacement) by moving on the three sides of a triangle, as depicted in the figure. Colored

[35] The Wino would be the "supersymmetric partner" of the *W*. Supersymmetry is an elegant unproved hypothesis whereby every known particle would have a so-far undiscovered more massive "super-partner."

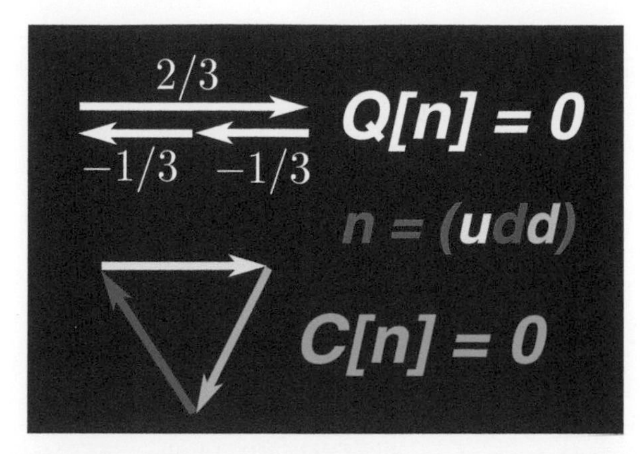

Figure 30 The electrical charges of the quarks in a neutron, $n = udd$, add up to zero, $Q[n] = 0$, the neutron is electrically neutral! The three colors of quarks add up in a manner analogous to vectors in a plane (the sides of the colored triangle). The neutron is "color-neutral," $C[n] = 0$, as well electrically neutral.

charges add up in a similar way and a neutron or a proton—made of three differently colored quarks—are also "color neutral"; their color is $C[n] = C[p] = 0$.

The proton and the neutron belong to a family of particles made of three quarks and called baryons, from the Greek for (relatively) "heavy." There are also bound states of a quark and an antiquark, called mesons, from the Greek for "not so" heavy. Examples of mesons are the pions, of which there are three. The π^+ is made of an up quark and an antidown one, $\pi^+ = (u\bar{d})$. The π^- is the antiparticle of the π^+, $\pi^- = (\bar{u}d)$. Finally, the π^0 is made of equal amounts of $u\bar{u}$ and $d\bar{d}$. Baryons and mesons are shown in Figure 31.

The consequences of the mathematical realization of the peculiar behavior of QCD charges are astonishing. A good old QED photon has no charge. When emitting or absorbing photons, an electron keeps its charge unchanged. In this QED and QCD differ. A blue quark may become red by the emission of a "blue-antired" gluon, as in the top of Figure 32. Gluons have color and may emit or absorb other gluons, as in the middle row of the figure, which has no analog in QED, since photons are neutral. A pair of gluons may also couple to another pair, as in the bottom row of the figure. So what?

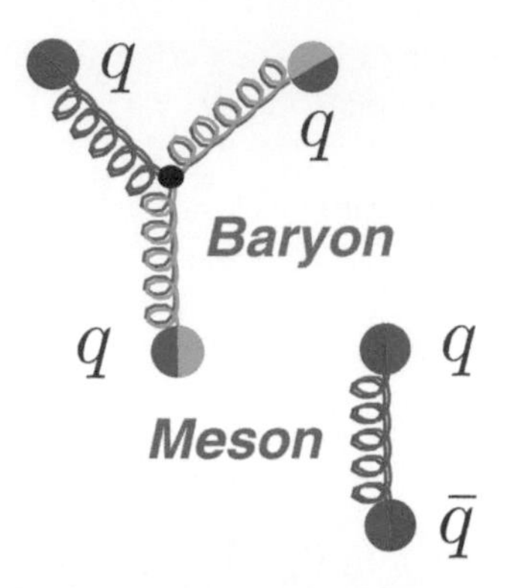

Figure 31 A *baryon*, such as a proton, is made of three quarks, labelled q. A *meson*, such as a pion, is made of a q and an antiquark, labelled $\bar{q}$. Both are glued by gluons. The color labels are such that a quark changes color when emitting a gluon, so that color "is conserved." The interactions between quarks and gluons are depicted in Figure 32.

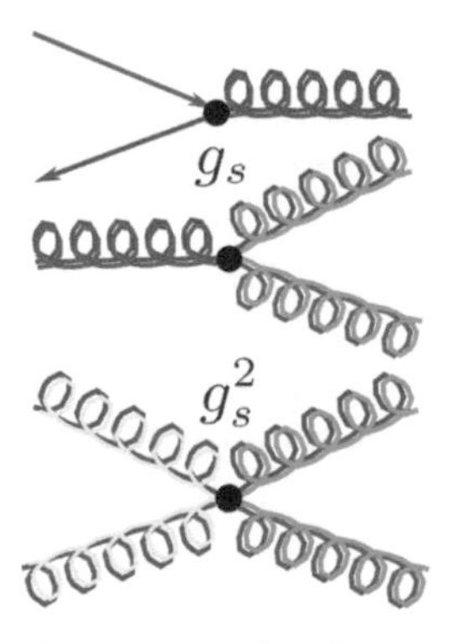

Figure 32 A colored quark emits a colored-anticolored gluon, to become a quark of a different color [top]. Similarly, a colored gluon may become a differently colored one by the emission of a gluon [middle]. Up to an explicit factor, quarks and gluons have the same colored charge, g_s, describing the size of their strong interactions. There is also a "four-gluon coupling," proportional to g_s^2 [bottom].

The confinement of quarks and gluons

Quarks and gluons are *observed* to be *confined*. That is, no (correct) experiment has ever been able to detect a *free* quark or gluon. A free quark (or antiquark) would have an extremely clear signature: A "fractional" charge of +2/3 or −1/3 (or −2/3 or 1/3). All other known particles have charges that are not fractional, −1 (+1) for the electron (proton), for instance.

Amongst the many experiments searching for free quarks, perhaps the most original one, performed by Peter Franken, was based on analyzing oysters. The liver of an oyster, it is stated, is one of the best filters anywhere. Every day it processes an amount of water some thousand times its weight. In so doing, it accumulates all sorts of peculiar substances in the oyster's growing shell. An atom or molecule containing one extra quark would have a fractional electrical charge and should have a very peculiar chemistry. It would presumably be caught by the oyster's liver. Alas, the experimentalists did not find any quarks. But, in the process of studying a barrel of oysters a day, they presumably gained some weight.

There are two reasons to conclude that the hydrogen atom is made of a proton and an electron. The first is that this "model" works, in the sense of correctly describing the states of hydrogen and the transitions between them, illustrated in Figure 27. That is 99% convincing. But there is a second reason, 101% convincing, so to speak. If, as in Figure 33, you break hydrogen by hitting it with a sufficiently energetic photon, it breaks into . . . a proton and an electron.

Similarly, a baryon is made of three quarks and a meson is made of a quark and an antiquark: The various states of such bound systems snugly coincide with the observed ones. But try to break a meson into its constituents, as in Figure 33. The energy of the photon is partially transferred to the gluons that bind the quarks and results in the creation of a quark/antiquark pair. The end result are two mesons, *not* a quark and an antiquark. *There are things that are made of pieces, but when you break them to pieces . . . they do not break into their pieces.* In short, this fascinating fact of Nature is often called *Quark Confinement*, though gluons are also confined.

So far we have referred to the electron charge as if it had a fixed value. In a sense, that is not quite correct. If one looks at an electron at shorter and shorter distances, the value of its "effective" charge increases.[36]

As it turns out, the fact that gluons are colored and interact directly with themselves makes an enormous difference between the behavior of

[36] The force between two interacting electrons at a distance r is not proportional to $1/r^2$ (as it would be if the product of their charges was a constant), it increases faster than $1/r^2$ as r decreases. This is the sense in which the "effective" charges—and the consequent strength of the forces between them—varies.

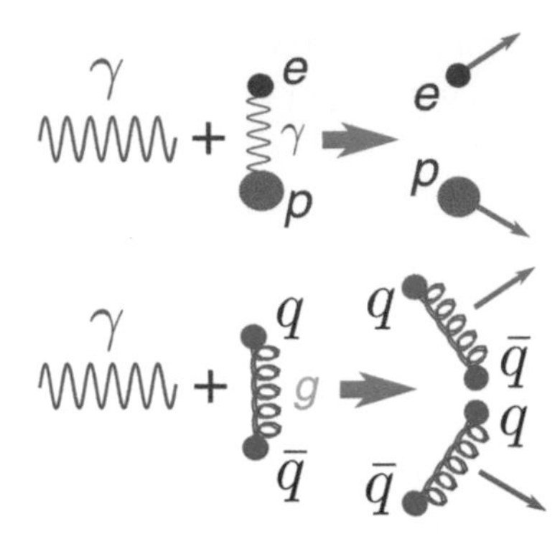

Figure 33 If you hit a hydrogen atom with an energetic photon (γ), it breaks into its constituents; an electron (e) and a proton (p). If you attempt to break a meson, made by a quark (q) and an antiquark ($\bar{q}$), it does not break into a q and a $\bar{q}$, but into a pair of mesons. Quarks cannot be isolated, they are *confined*.

electric and colored "effective" charges: The latter *decrease* at increasingly shorter distances. This *prediction* of the theory of QCD is called *asymptotic freedom*. An apparent alternative way of saying the very same thing is that the forces between the colored objects binding into an uncolored one increase with no limit as the distance between them increases and increases. This would apply to quark/antiquark pairs in a meson and to the three quarks in a baryon: It would seem to be the theoretical proof of quark confinement. It is not.

As the forces between separating quarks become stronger and stronger, it becomes increasingly difficult for QCD theory—in its current state of development—to deal with them. The methods used to prove asymptotic freedom—for quarks and gluons in a domain where their interactions are relatively weak—are no longer valid when the interactions become strong. Thus, there is no clearcut totally satisfactory ship-shape proof that "QCD confines." Instead, there is a prize of one million dollars awaiting the first person to give a theoretical proof of confinement... to the satisfaction of the jury.[37]

We have seen in Chapter 2 how touchy the question of priority in authorship may be, particularly in cases where results by different people are obtained almost simultaneously. The proofs of asymptotic

[37] *The Seven Millennium Problems.* The Clay Mathematics Institute of Cambridge, Massachusetts (CMI). http://claymath.org/millennium-problems/millennium-prize-problems.

Figure 34 The Harvard and Princeton shields. The ambitious meaning of Harvard's motto is transparent. I am told by expert latinists that the Princeton motto means "God went to Princeton." https://en.wikipedia.org/wiki/File:Harvard shield-University.png Author unknown (1644). Current design by Pierre la Rose (1895). https://commons.wikimedia.org/wiki/File:Princeton-shieldlarge.png by huggo (Own work).

freedom by David Politzer (then at Harvard) and David Gross and Frank Wilczek (then at Princeton) were received by the journal *Physical Review Letters* with six days between them. This is but one example of the time-honored competition—not only in rowing—between these two American universities. Their crimson and orange shields are shown and commented on in Figure 34. You have correctly guessed: at the time I was at Harvard Gross, Politzer, and Wilczek received the 2004 Nobel Prize "for the discovery of asymptotic freedom in the theory of the strong interaction."

8

Everything . . . So Far

Time to summarize the list of elementary particles we have encountered so far, which appears in Figure 35, divided into **fermions** and *bosons*. Here, as in politics, those on the left column more often than not act as *constituents*, while those of the right act as *forces*.

The precise meaning of *spin* we shall discuss anon. For now, inaccurately speaking, the spin is a measure of how an object "rotates around itself." The fermions in the table of Figure 35 have spin 1/2. The listed bosons have integer spin, all of them 1, except for the graviton, whose spin is 2. The electron and the neutrino belong to a family of particles that have no color (no strong interactions). They are called **leptons**, from the Greek for *light-weight*. To add to the increasing absurdity of the accumulating names of the elementary particles and forces . . . we shall soon encounter **heavy leptons**. A sensible denomination, for once, is **hadrons**, from the Greek root for "strong." They are the particles made of quarks that have strong interactions.

FERMIONS	BOSONS
up quark, u	Photon, γ
down quark, d	Gluons, g
electron, e	I.V.Bs, W, Z
neutrino, ν	Graviton, G

Figure 35 Fermions and bosons. *I.V.Bs* stands for **Intermediate Vector Bosons**, the carriers of the *weak* interactions. The horizontal bar divides the quarks from the *leptons*, namely the electron and the neutrino.

Enjoy Our Universe: You Have No Other Choice. Alvaro De Rújula.

DOI: 10.1093/oso/9780198817802.001.0001

9

A Parenthesis on Quantum Mechanics

Quantum mechanics is a fact of life, in the sense that, as far as we know, it is part of the basic functioning of Nature. *Quantum* refers to several of its tenets. One is that particles, such as the photons constituting the electromagnetic radiation, come in "packets," or *quanta*, the observation that earned Einstein his Nobel Prize. The photons' energy, E, and the frequency of the radiation, ν, are related: $E = h\,\nu$, with h the *Planck constant*. In *natural units,* physicists not only set $c = 1$ but also $\hbar = h/(2\pi) = 1$. "Quantum" also refers to the fact that some other things, such as the energy levels of an atom, are not arbitrary. They are *quantized*, as opposed to *continuous*; that is, they cannot be dialed to intermediate or capricious values.

Classical mechanics refers to the description of Nature in the "approximation" in which quantum effects are ignored, quite often a very good one. The energy levels of a planet circling the Sun, for instance, are also quantized. But, unlike for the much smaller atoms, the difference between two neighboring planetary energy levels is not relevant, being completely negligible relative to their individual values.[38]

Classical mechanics is deterministic. Given the position and momentum of a particle it is possible to predict its future trajectory. Quantum mechanics has an intrinsic uncertainty. Position and momentum (or velocity) cannot be simultaneously measured with arbitrary precision. This *Heisenberg's uncertainty principle*[39] is illustrated in Figure 36. Nor can one predict the future, except in the sense of the relative probability of distinct outcomes.

The *two slit experiment* is a notorious example of quantum-mechanical uncertainty. Consider the apparatus depicted in Figure 37. A source emits photons that impinge on a non-transparent screen with two

[38] Quantum states are also affected by interactions with external or internal agents, losing their "quantum coherence" and identity. A planet emits and absorbs light, does not travel in a perfect vacuum, etc. It is not in an unperturbed quantum state.

[39] The uncertainty in position, Δx, and in momentum, Δp, are such that their product cannot be smaller than $\hbar/2$.

Enjoy Our Universe: You Have No Other Choice. Alvaro De Rújula.

DOI: 10.1093/oso/9780198817802.001.0001

Figure 36 The Heisenberg uncertainty principle. The location and momentum of an object cannot be simultaneously determined with arbitrarily high precision. In the car, Werner Heisenberg in his youth (1926). Wikipedia photo from Friedrich Hund.

open holes. On the other side there is a CCD camera sensor precise enough to record the position at which every single photon arrives. If the wavelength of the photons is much smaller than the slit's apertures, the result is *ballistic*; the photons behave like bullets or classical particles; they reach the screen following the continuation of the straight-line trajectories from the source through one slit or the other. The number of photons seen by the detector screen at the different positions is described by the blue areas in the figure, the illuminated images of the slits.

If the wavelength of the photons is comparable to, or larger than, the slit's apertures, they behave like waves. The two slits act as secondary sources of waves that interfere, resulting in a pattern of arrival photon numbers such as the red one in the figure. The photon wave from the source has "gone through the two slits." This is not so surprising, one can in a bath-tub make a similar experiment with water waves.

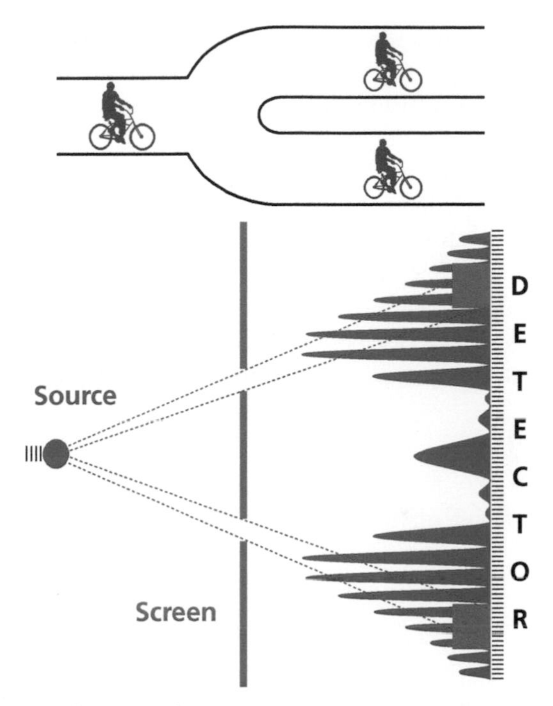

Figure 37 A zen dictum and its quantum-mechanical realization. A source of light illuminates a screen with two slits. Short wavelength photons form the blue image. Its vertical extent tells us where they hit, the (blue) horizontal one tells us the number of photons in each of the (black) detector elements. Longer wavelength photons—even if emitted and recorded one by one—build up the red "interference pattern."

And now... the stunning wonder, reminiscent of Zen advice: *If you find a fork on your way, take it!* Suppose that, continuing to study the wavy long-wavelength photons, one emits them from the source *one by one*. The detector still records their arrival position, and even beeps and blinks for you to know that a photon did arrive and where. A single photon may be expected to go through one slit *or* the other, and be recorded at one or the other of the (blue) positions of the images of the slits. Lo and behold, this is not the observed result. The accumulated single photons build up the red image of arrival positions. This interference pattern implies that every *single* photon *went through both slits*.[40]

[40] To be more precise, it is the "wave function" or the quantum-mechanical "path" of a single photon that "went" through both slits.

One cannot predict where a particular single photon would end up. The interference pattern is the *probability distribution* of their arrival positions. That is all that in quantum mechanics (and in reality) one can predict. And this is not just a thought experiment!

Not only photons, but also electrons, atoms, or even fairly large molecules ("buckyballs" made of 60 carbon atoms) have been observed to pass simultaneously though two holes separated by a distance much larger than their size. Current technology does not allow one to perform the experiment (with success) with cats as projectiles. Herding cats was never easy, in this case because it is difficult to "prepare" and keep them in quantum states as well defined as those of single photons or molecules.

10

Understanding Relativity and Quantum Mechanics

To *really* understand something, say tennis, relativity, or quantum mechanics, you have to be a practitioner, in which case your understanding reaches the point where you "feel it in your bones." There is nothing more frustrating than having a tennis pro tell you: "It is so simple, just do like this.... " I shall not do anything of the sort.

Feeling relativity in one's bones is made difficult by our lack of experience with velocities comparable to that of light. At rest, relative to the famous Hawaiian-born wrestler Akebono, you would see him as on the left of Figure 38, while if moving sideways at half the speed of light

Figure 38 Akebono seen by an observer also at rest on the ring [left] or passing by at a relativistically fast sideways motion [right]. Adapted from a panoramio.com photo by Gio la Gamb. http://www.panoramio.com/photo/34205046.

Enjoy Our Universe: You Have No Other Choice. Alvaro De Rújula.

DOI: 10.1093/oso/9780198817802.001.0001

you would see him *relativistically rotated,* as on the right.[41] Confronting this guy one may, indeed, want to run away that fast.

Feeling quantum mechanics in one's bones is more difficult, even if the chemistry that builds them up is quantum-mechanical. Again, the problem is one of "everyday inexperience." Yet, it would be easy to have children play with the two-slit experiment and grow up accustomed to quantum uncertainty. On the other hand, sending them to take pictures of relativistically rotated or contracted objects would be unaffordable.

[41] To be precise: The finite speed of the light reaching your eyes makes Akebono seem rotated, as opposed to contracted. A sideways moving seal stamping (simultaneously in your reference system) a paper you are holding would leave a contracted mark.

11

Spin, Statistics, Supernovae, Neutron Stars, and Black Holes

Consider, as in Figure 39, two balls of mass M at the ends of a rod, hanging from the ceiling and rotating around the vertical suspension. In an ideal case in which the balls are much more massive than the rest of the "apparatus," one would observe that, if the balls move further away as the system rotates (L increases), their rotational velocity, v, decreases. With balls, as opposed to a ballerina, it would be fairly easy to discover a new *conservation law:* The **angular momentum**, $M\ v \times L$, remains fixed: it is "conserved."

Angular momentum also comes in discrete or quantized units. Its possible values are *integer* (0, 1, 2...) or *half integer* (1/2, 3/2, 5/2...) multiples of $\hbar$. Thus, in natural units, the "intrinsic" angular momentum (the *spin*) can take the values 0, 1/2, 1, 3/2, 2...The "quantization" of

Figure 39 When extending her arms and legs, a ballerina rotates less fast. So do two balls as the distance between them, L, is increased. The product $v \times L$, however, is unchanged (v is the ball's velocity of rotation).

Enjoy Our Universe: You Have No Other Choice. Alvaro De Rújula.

DOI: 10.1093/oso/9780198817802.001.0001

angular momentum[42] underlies a definition of spin more precise than the usual sloppy one ("the amount by which something rotates about itself"). Spin determines the way an object behaves under rotations; that is, either you rotate it and see what happens, or you leave it alone and, while looking at it, you perform the opposite rotation yourself.

Elementary or composite particles obey two and only two *statistics*: The description of how they can be *packed*. If you attempt to fill a box with photons you will not succeed; there is always room for extra ones. More surprisingly, the process of adding photons (in the same quantum state as the previous ones) becomes easier and easier as they accumulate, this being the principle underlying the way a *laser* works. Contrariwise, if you fill the box with electrons, or with neutrons, you cannot put more than one in the same state of energy and spin: The box gets "filled," see Figure 40. Ultimately, this is the reason why an object, such as you or a planet, is "solid" or why the force of gravity does not collapse a star.

In some very massive stars, however, gravity eventually wins. Their central part, or *core*, is sustained by the electrons, that cannot be more packed. But there comes a point when the core, millions of years old, collapses in a fraction of a second. The electrons are "eaten" by protons, to become neutrons and neutrinos ($ep \to n\nu$). The weakly interacting neutrinos easily manage to escape the star. The core collapse stops when the neutrons get squeezed into a *neutron star*, supported now by the fact that neutrons cannot be happily packed. For a very very massive star, not even the neutrons manage to stop the core's collapse and the result is a *black hole*, a possible "solution" to Einstein's equations in Figure 1. In these processes, the matter of the parent star outside its core is violently expelled in a way that, admittedly, we do not yet understand in fully satisfactory detail. The result is a supernova; observable as a very luminous transient light in the sky, the death pang of a star.[43]

Up to now I have referred to *black holes* with no ado, abusing the fact that they have become tremendously familiar, at least in the sense of having turned into a mundane expression. My first encounter with them was in a typically horrifying anonymous childrens' tale called *El*

[42] A normal-sized ballerina rotating once a second would have an angular momentum of some $10^{34}\hbar$. Adding or subtracting one quantum of angular momentum (plus or minus $1\hbar$) has a negligible effect. That is why quantum mechanics was not discovered by dancers.

[43] There is another supernova "type" which we understand even less well.

Figure 40 Particle "statistics": Fermions cannot be comfortably accumulated [top], while the ghostly bosons can [bottom]. King Penguins photo by DocAC in flickr.com. Empty wagon photo from pixabay.com.

Figure 41 If you approach a black hole beyond a surface called the *event horizon* you will never be able to backtrack. M and R are the mass and radius of the hole, G is Newton's constant. The inequality (in $c = 1$ units) is the condition for "it" to be a (neutral non-rotating) black hole. Black hole: Wikipedia image by Alain r.

castillo de irás y no volverás ("The castle to which you will go, but from which you will not return"). That is precisely what they are, see Figure 41.

The spin-statistics theorem

One of the two deepest and most unfathomable consequences of the marriage of relativity and quantum mechanics is the *spin-statistics theorem.* The theorem states that all fermions (all particles with half-integer spin) are of the kind that would "fill a box," while all bosons (all particles with integer spin) are of the other kind, for which additional ones are always welcome to the club. Thus, the spin of a given object determines how ensembles of identical ones behave (their "statistics").

The spin-statistics theorem describes a weird fact of Nature, mathematically well understood—though with considerable toil—and extremely difficult to (honestly) "prove" with only words. One reason for that is that the theorem involves two very peculiar quantum-mechanical facts:

The state of a particle is described by its "wave function," whose square at a given point is the probability of the particle being there. This wavy property underlies the "zen" behavior of the photons or electrons in Figure 37. Rotate a boson by 360° and you get the same wave function it had before. Of course! you say. But, turn a fermion by the same amount and you get *minus* the original wave function. This is a serious blow to ones's "classical" conviction that rotating something by a full circle is entirely ineffective, useless.

The second oddity concerns the wave function of a pair of *identical* particles.[44] Do a *permutation:* Exchange the particle's positions so that one is placed where the other one was, and vice versa. The pair's wave function does not change if they are bosons, but it changes sign for fermions. A blow to one's "classical" view of *identity.* If Figure 25 pictured the wave function of identical brothers—who are fermions—and you exchanged their position, what would you get? *Minus* the brothers.

Treating pairs of objects that change sign the way fermions do requires introducing mathematical thingamajigs for which a times b does not equal b times a. An excellent reason not to do any more algebra here.

The effects of a 360° rotation and a permutation are the same: None for bosons, a change of sign for fermions. This is not a coincidence but the basis of the spin-statistics connection. The theorem indeed relates spin (the behavior of an object under rotations) and statistics (the way an ensemble of identical objects behaves).

[44] Located at places that could have communicated at a speed equal or smaller than c. This is how relativity also plays a role in the theorem.

12

Parallel Realms

We encountered in Figure 35 a preliminary list of "everything." The list is incomplete, but it is in a sense the list of all things inescapably required—for you to be reading this, that is.

We have known since 1936 that the electron has a *cousin,* the *muon,* μ; an ingredient of Nature that nobody "had ordered." Here and in what follows, *cousins* describes the relation between particles that are essentially identical but for their mass. Like the electron, whose identity card is in Figure 24, the muon has four basic properties. It has the same charge and spin as the electron, but it is some 210 times more massive. With disrespect for the language of the Greek forefathers of science, the muon is dubbed a *heavy lepton*. It is the first of the series of novel characters starring in the cast of Figure 42.

The massive nature of the muon results in it being unstable, muons decay in much the same way as the down quark in Figure 29. Their predicted and observed average lifetime is 2.2 millionths of a second. By now we know that μ decay results in two *different* types of neutrinos, $\mu \rightarrow e\, \nu_\mu \bar{\nu}_e$. Here $\bar{\nu}_e$ is the "electron's" antineutrino, which we did previously encounter, and ν_μ is the *muon's neutrino*, the brother of the muon.[45]

Starting in 1947, a series of new hadrons was discovered (recall that hadrons are strongly interacting particles, made of quarks). Their properties seemed rather peculiar, to the point that they were dubbed *strange particles*. By now we know they are pretty normal but for the fact that they contain one or more *strange quarks*, s, or antiquarks, $\bar{s}$. One of the first strange particles discovered was *the Lambda*, similar to a neutron (udd) but with one of the d quarks substituted for an s, $\Lambda = (uds)$. Another example are *the kaons*, pions with an "ordinary" quark (u or d) substituted for a

[45] This "brotherhood" means that in "charged" weak interactions (mediated by Ws) the neutrinos appearing in association with an electron or with a muon are different. And ν_e and ν_μ are *defined* as the entities associated in this way with es and μs, respectively.

Enjoy Our Universe: You Have No Other Choice. Alvaro De Rújula.

DOI: 10.1093/oso/9780198817802.001.0001

FERMIONS		NEWEST BOSON
OLD	NEWER	
$\binom{u}{d}$	$\binom{c}{s}$ $\binom{t}{b}$	H
$\binom{e}{\nu_e}$	$\binom{\mu}{\nu_\mu}$ $\binom{\tau}{\nu_\tau}$	

Figure 42 The list of all known elementary *fermions* and the most recently discovered elementary boson "the Higgs" (H). Reading vertically, a complete column of fermions is called a *family*. Its two first entries (u and d in the first family) are quarks, the other two (e and ν_e) are *leptons*. Quarks have color (strong interactions), electrons and neutrinos do not.

strange one. Putting order in this zoo was what *the quark model* eventually managed to do.

The s quark has the same properties as the d quark but for its mass, which is greater. The d and s quarks are two cousins, like the e and μ pair.[46] The s quark has sufficient mass to enjoy the luxury of decaying in various ways: $s \to ud\bar{u}$, $s \to ue\bar{\nu}_e$, $s \to u\mu\bar{\nu}_\mu$.

For a long time, the list of leptons, $\{(e, \nu_e); (\mu, \nu_\mu)\}$, and the one of quarks $\{(u, d); s\}$, seemed strangely asymmetrical. At a time when practically nobody believed that quarks were real, a handful of physicists, particularly Sheldon Glashow and his collaborators, "already knew" that leptons and quarks were analogous and that a fourth quark, termed *charmed* and denoted c, had to exist. Its existence would resolve old inconsistencies between the theory of weak interactions and the relevant observations, as if by magic. Not unscholarly, charm also means "spell."

The November Revolution

In November 1974, a scientific revolution took place, the first particles containing c quarks were discovered. But Nature decided to add some

[46] I could not add "like ν_e and ν_μ" because, as defined in footnote 45, these particles are mixtures of neutrinos of different masses. All other particles in Figure 42 are defined by having a specific mass.

Figure 43 One of the two first charmed particles to be discovered: (*udc*) in the iconic notation of Figure 44.

spice to the story and these particles turned out to be bound states of a charmed and an anticharmed quark ($c\bar{c}$). Far from being obvious, their *charm* was discreet: A single c decays in a tell-tale characteristic manner, but a $c\bar{c}$ pair annihilates into "uninteresting stuff." The existence of truly charmed particles, containing just one charmed quark, had to wait two more years to be experimentally demonstrated. The first ones to be discovered were the bound states (*uuc*) and (*ucd*), the second of which is shown in Figure 43.

Positronium is the simplest of artificial atoms. It consists of an electron and a positron, $e\bar{e}$, and it is an optimal "laboratory" for the study of QED. Its ground state decays fast into three or two photons, depending on whether the spins of the $e\bar{e}$ pair are aligned or anti-aligned. The $c\bar{c}$ particle discovered in 1974 is analogous to the second of the mentioned $e\bar{e}$ states (the other one also exists, but it took an extra couple of years to find it). These objects were immediately dubbed *Charmonium*, which does not mean that, for starters, more than a few people proposed this interpretation, or believed in it.

The first two years after the November Revolution the cows were lean. This means that the available experimental data had indirect evidence for the presence of charmed quarks and of the τ lepton listed in the table of Figure 42. The evidence, extracted by theorists with the help of the then not generally accepted theory of QCD, was totally convincing... for these theorists! Those were the good old times in which—as they should and as most of them no longer do—experimentalists attempted to prove theories wrong, not right. That had the enormous benefit of making experimental evidence *in favor* of something much stronger: "In spite of our streneous efforts to achieve the contrary, we have discovered what we did not want to" is an eminently convincing argument.

Finally, in 1976, a collection of charmonium states with properties very similar to the states of positronium (but for the substitution of QED for QCD) were finally discovered. This, and the discovery and properties of the (*uuc*), (*ucd*), and other charmed particles, finally convinced the "community" at large of the reality of quarks. The building blocks of the so-called Standard Model of particle physics were almost all established. By then, four quarks were known, the ones depicted in Figure 44.

The third family

In the table of Figure 42 three *families* are shown, each consisting of a pair of quarks and a pair of leptons. The first member of the third family to be discovered was the *tau lepton*, τ, identical to the electron or the muon, but for its being more massive. It took its brother, the *tau neutrino* (ν_τ), years to be definitely established; the weakly interacting neutrinos are hard to observe and study, recall Figures 21 and 22.

The most massive pair of quarks, *t* and *b* in the table of Figure 42, ought to have had their names based on *tenderness* and *beauty*, in which case it would have been a pleasure to add them to Figure 44. This proposal did partly catch on, in the case of the "beauty" of the *b* quark. But the prevailing nomenclature is *top* and *bottom*, which explains why it might have been considered uncouth to add them to the figure. Once again, the beauty and tenderness of these quarks is but a name. They are almost identical to the other quarks lying in Figure 42 in the same

Figure 44 The four quarks known to exist in 1974, in their three arbitrary "colors," swiftly recognizable by Spaniards.

horizontal row to which each of them belongs. All that tells them apart is their being more massive than the ones to their left.

Particles containing *t* and *b* quarks have been discovered. The top quark was the harder one to make, since its mass is record-breaking for an elementary particle: 340,000 times the mass of an electron. For completeness, let me refer to yet another concept with a silly name.

Each of the fermions listed in Figure 42, particularly the quarks, is called a different **flavor**. As if we could imagine the flavor resulting from topping ice-creams with top quarks.

Are there other families waiting to be discovered? The families in Figure 42 all contain neutrinos, which are much lighter than the other particles. If new families are defined as sharing this property—and the same interactions as the known families—the answer is no: We have already filled the tribe. The reason is that the *Z* boson would decay, among other things, into the neutrino–antineutrino pairs of the new families, and that is observationally excluded (an extra decay "channel" would make *Z*s have a shorter mean lifetime than the predicted and observed ones, which agree—if there are only three neutrino types).

I have stated in passing that the charmed quark was "required" by the theory, and in that sense it was not a complete surprise. The same is true for the pair of heaviest quarks, *b* and *t*; we "needed" them. What do these "requirements" and "needs" mean? This deserves its own Chapter: the next.

13

A Parenthesis on "R^2QFTs" ★★

The acronym in the above title is meant to hide for a second its undoubtedly frightening meaning: **Renormalizable Relativistic Quantum Field Theories**. These are the most solid and predictive tools in our understanding of elementary particles and their interactions. The *Standard Model* of particle physics is a R^2QFT, an honest-to-goodness theory and not a mere "model." Contrariwise, we do not have a satisfactory R^2QFT of gravity, yet.

In practice, "renormalizable" means that these theories have a few parameters—such as the mass and charge of the electron—which have to be taken from observation, they are not a prediction. But given these input parameters, some toil, and various levels of precision, all other "facts" are predictable; for example, other properties of the electron and its electromagnetic and weak interactions.

An electron is a little magnet that can be oriented in a magnetic field, see Figure 45. In a ferromagnet, the spins of a good fraction of

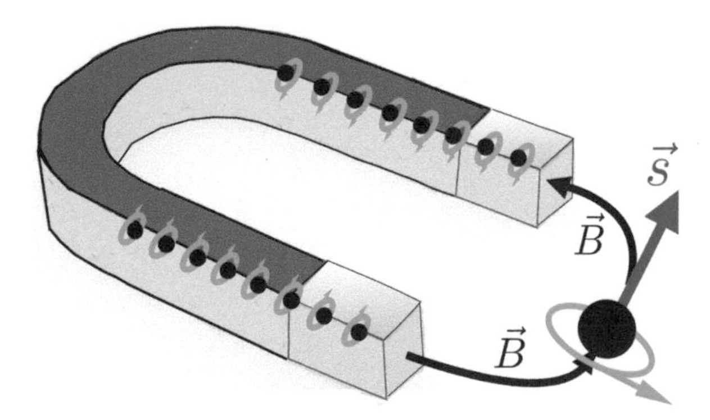

Figure 45 The spins of many electrons in a ferromagnet are aligned. They generate a magnetic field, $\vec{B}$. The field orients the spin $\vec{s}$ of an external electron in its direction.

Enjoy Our Universe: You Have No Other Choice. Alvaro De Rújula.

DOI: 10.1093/oso/ 9780198817802.001.0001

its electrons are all aligned, creating the observed collective magnetic field. An external electron placed in this field would orient its spin along it.

In natural units that include $m_e = 1$, and to a first approximation, the value of an electron's *magnetic moment* (the "strength" of its magnetism, aligned with the spin) is predicted to be $g=2$. Physicists like to talk in terms of the *anomaly* $a=g-2$. The anomaly vanishes ($a=0$) for the simplest interaction, pictured in Figure 46(a), of an electron and the photon that represents the magnetic field here. But the interaction can involve another photon, emitted and reabsorbed by the electron, as in Figure 46(b). This modifies the prediction for a, which no longer vanishes. The modification is called of *second order* in the charge e of the electron because the extra photon interacts twice with electrons—as in the dark-dotted places labeled e in Figure 46(b)—and the result is proportional to e^2, the square of the electron's charge.

To second order in the charge e, the result is $a=\alpha/(2\pi)$, with $\alpha=e^2/(4\pi) \approx 1/137$. This is the first of a series of corrections[47] in an "expansion" in successive powers of the small quantity α. To *fourth order*, there are seven diagrams, one of which is shown in Figure 46(c), wherein the *loop* represents an electron-positron pair. To *sixth order* the seventy two *Feynman diagrams* are those of Figure 46(d).

It took theorists sixty eight years of effort to proceed from the first approximation ($g=2$, $a=0$) to the complete analytical result up to the sixth order, shown for fun and illustration in Figure 47—the reader is not required to check that it is correct.[48] Every order results in a much smaller correction than the previous one. In April 2017, Stefano Laporta published the eighth-order analytical result and specified the numerical value of c_4—for which there is no room in Figure 47—to 1100 significant digits! The calculation of the tenth order is in progress. Obviously, I am not drawing the 891 (or 12,672) eighth order (or tenth order) Feynman diagrams here.

The prediction for the anomaly is the sum of the contributions to various orders. The current experimental observation is $a=1{,}159{,}652{,}188 \times 10^{-12}$, with an error of four parts per billion. Theory

[47] Such a calculation in successive approximations is called "perturbative," from the original example of predicting the orbits of planets as if dominated by the Sun's gravity, "perturbing" them by the effect of the largest planet (Saturn) and so on.

[48] The diagrams of the figure contain only electrons and photons. There are subdominant ones involving other particles.

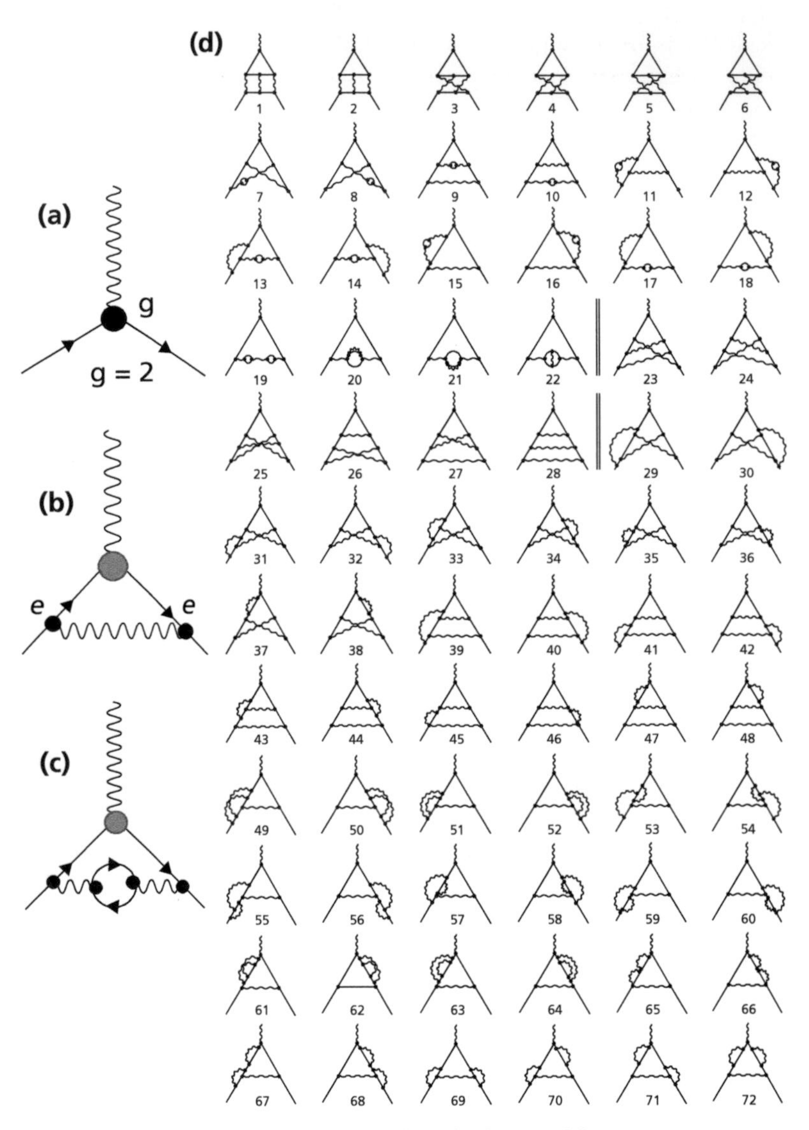

Figure 46 *Feynman diagrams* relevant to the calculation of the magnetic moment of the electron, g. (a) The leading prediction: $g=2$. (b) The first correction, with one photon being exchanged, contains two extra electron-photon interactions and is of order $\alpha\,(e^2)$. (c) One of the seven diagrams to order $\alpha^2\,(e^4)$. (d) All of the seventy-two diagrams to order $\alpha^3\,(e^6)$ [B.E. Lautrup, A. Peterman & E. de Rafael. Phys. Rept. 3 (1972) 193-260]. Each perturbative correction is much smaller than the previous one.

$$g = 2 + a_e \qquad 2 : \text{Dirac } 1928$$

$$\boxed{a_e = c_1 \left(\frac{\alpha}{\pi}\right) + c_2 \left(\frac{\alpha}{\pi}\right)^2 + c_3 \left(\frac{\alpha}{\pi}\right)^3 + \ldots}$$

$$c_1 = 1/2 \ \text{Schwinger } 1948$$

$$c_2 = \frac{197}{144} + \frac{\pi^2}{12} - \frac{1}{2}\pi^2 \ln 2 + \frac{3}{4}\xi(3)$$
$$= -0.328\,478\,965\,579$$

Petermann 1957, Sommerfeld 1958

$$a_e|_{\text{exp}} = 1\,159\,652\,188.4\,(4.3) \times 10^{-12}\ (4\text{ppb})$$

$$c_3 = \frac{83}{72}\pi^2\xi(3) - \frac{215}{24}\xi(5) - \frac{239}{2160}\pi^4$$
$$+\frac{139}{18}\xi(3) - \frac{298}{9}\pi^2 \ln 2 + \frac{17101}{810}\pi^2 + \frac{28259}{5184}$$
$$+\frac{100}{3}\left[a_4 + \frac{1}{24}\ln^4 2 - \frac{1}{24}\pi^2 \ln^2 2\right]$$
$$= 1.181\,241\,456\ldots \ \text{Laporta and Remiddi } 1996$$

$$\boxed{a_e|_{\text{th}} - a_e|_{\text{exp}} = (32 \pm 28) \times 10^{-12}}$$

Figure 47 The analytical result for the electron's *anomaly*, in terms of the measured value of α and numbers: 2, π, $\xi(3)$, and so on. I promised not to show formulas, this is a *super-after-truth*. However, this particular one is true. S. Laporta & E. Remiddi, *Phys. Lett.* B379 (1996) 283–91, and references therein.

and experiment agree to better than three parts per billion in the value of *a*, corresponding to one part in ten trillion in the value of *g*. Quite a success for QED, the quintessential R²QFT. There are things we seem to understand with some precision! The electron's anomaly is but an example. The understanding of the Periodic Table of the elements, the levels and transitions in atoms, the emission and absorption of electromagnetic waves, all of chemistry, the properties of materials, and so on... are all consequences of QED.

Recent experiments have obtained a result that slightly deviates from the theoretical expectation. This "tension" is nothing to write home about, which does not stop scores of theorists from writing papers on possible more or less revolutionary explanations. Just in case currently starting experiments confirm the discrepancy.

Figure 48 Imperator Caesar Augustus pronounces a dictatorial decree. Wikimedia image by Till Niermann.

"Relativistic" and "Quantum" refer to theories that obey the laws of (special) relativity and of quantum mechanics. Finally, a *field*, of which there is one per particle, is a vastly powerful entity. It describes particles and their antiparticles and allows one to specify the interactions between different fields that determine what happens in the processes in which they are involved, including the possible creation or destruction of particles.

In a sense, R^2QFTs are like dictatorships, see Figure 48. *Everything that is allowed is compulsory, and all things not explicitly authorized are forbidden.* In the Standard model, as we have mentioned, certain decays of the strange quark—such as $s \to du\bar{u}$—which are not observed, *must* happen. Unless there exists—or existed[49] at the time it was postulated—a charmed quark. If so, the unseen decays are forbidden.

If you see a video of an actual collision of two billiard balls you may conclude that it was a real video and not a computer simulation. If you look at the same video backward, you reach the same conclusion.

[49] Recall from Chapter 1 how Goddess adds recipes to her cook book.

That is because, to a very good approximation, the fundamental laws of Nature are *time-reversal invariant.*[50] But it has been observed for over half a century that Nature slightly *violates* time reversal invariance. The Standard Model makes time-reversal invariance mandatory, unless the *b* and *t* quarks also exist, in which case the invariance is violated at the observed level. That is how the *b* and *t* quarks were predicted.

Proving that a theory which correctly describes Nature is a R²QFT is highly non-trivial. To the point of deserving a Nobel Prize. Awarded in 1965 to Sin-Itiro Tomonaga, Julian Schwinger, and Richard (Dick) Feynman for QED and in 1999 to Gerardus (Gerard) 't Hooft and Martinus (Tini) Veltman for QCD. Cautiously, I have ordered them as in Nobel's official citations. Odin only knows how the Nobel Committee chooses to order its Laureates. If it always was anti-alphabetically we would have learned where to place a *'t.*

The Mother of all Concepts

Particles and their creation and destruction, forces and their action at a distance, waves... All these concepts that we have discussed are described by a single unified entity: A *Relativistic Quantum* Field (RQF). It is as if this was the *Mother of all Concepts*, the only word in the language used to successfully describe Nature at a fundamental level.[51] A RQF is a local function of space and time. In its static version (no time dependence) it describes forces: Actions at a distance. In its collective dynamic version an RQF describes waves. Its quantum aspect describes individual particles that, in a demiurgic facet, can be created and destroyed.

[50] Alas, for a complicated process, such as the spilling of a glass of wine on one's lap, time-reversal seems to be impossible. But that is because of the utter difficulty of reversing the process all the way down to the atomic level.

[51] Other words, such as *(super-)string* have not, so far, resulted in testable predictions.

14

The Unification of Forces

We have discussed various *unifications* of two or more concepts "for the price of one," crucial steps in the hoped-for simplicity of our understanding of the basic elements of Nature. The progressive unifications of the apparently distinct basic forces are a particularly enjoyable example.

At least two fathers of relativity and quantum mechanics—Einstein and Heisenberg—spent many years attempting, with no success, to find a *Unified Theory of all fundamental forces.* One of the reasons why they failed is that they did not know about, or paid little attention to, two fundamental forces that do exist: The weak and the strong ones. To date, nobody has succeeded in this grand-unifying challenge. Who knows whether we are still ignorant of some hypothetical *Fifth Force?*

In 1831, Michael Faraday, presumably the best experimental physicist of all time, *unified electricity* by concluding that "frictional, galvanic, voltaic, magnetic and thermal electricity are one and the same" (Read by Faraday before the Royal Society on 10 and 17 Jan. 1833.) and discovered his *law of induction*, describing how a time-varying magnetic field induces an electrical current. The complete classical theory unifying electricity and magnetism was published by James Clerk Maxwell in 1861.[52] Maxwell's equations also *unified light and electromagnetism* in the sense of interpreting light as an electromagnetic wave and predicting similar waves of different frequencies, most notably radio waves. These were discovered by Heinrich Hertz on November 11, 1886, see Figure 49.

In the 1840s, Faraday suffered a mental breakdown, perhaps due to chemical poisoning or obsessive work. In the 1850s, very much like Einstein but with the extreme opposite—solely experimental—approach, he set to measure the induction of electricity by a body decelerating in the Earth's gravitational field. In Figure 50 we see him trying this first step in the unification of gravity and electromagnetism. He failed, like everybody else to this day. With characteristic honesty,

[52] The theory was not quantum mechanical but, amazingly, was in agreement with—and anticipated—special relativity.

Enjoy Our Universe: You Have No Other Choice. Alvaro De Rújula.

DOI: 10.1093/oso/9780198817802.001.0001

Figure 49 Top: The variously shaped antennae used by Hertz, and his radio-emitter, Volker Springel. ©Max-Planck-Gesellschaft zur Förderung der Wissenschaften e.V., Munich. All rights reserved. Bottom: A photo of Hertz's table-top experiment, taken by himself.

Faraday concluded: "Here end my trials for the present. The results are negative. They do not shake my strong feeling of a relation between gravity and electricity, though they give no proof that such a relation exists."

A century after Maxwell's unification, Sheldon Glashow, Steve Weinberg and Abdus Salam (GW&S) succeeded in unifying quantum

Figure 50 Faraday's failed experiment on what would have been his second law of induction (of electricity by gravity). The (super?) string may have been premonitory. Many of Faraday's instruments are preserved at the Royal Institution in Mayfair, London.

electrodynamics and the weak interactions into the *Electroweak Theory* that, added to QCD, constitutes the Standard Model.[53] In the Electroweak Theory, the photons mediating the electrodynamic

[53] My own significant contribution to this subject was to photocopy Glashow's doctoral thesis at the Harvard Library and send it to the secretary of the Physics Nobel Prize Committee, upon request (of the latter) and under conditions of absolute secrecy.

interactions and the $W^{\pm}$ and Z^0 *intermediate vector bosons* carrying the weak interactions are not independent entities. Like the sides of a dice, they can be "rotated" into one another, but in a specific mathematical "space" that is not the "ordinary" one we live in.

An electron can interact with a proton: They are both charged. Thus, they can exchange a photon and collide or "scatter" off each other, $ep \to ep$. In such a process the observer does not "see" the photon, which is not part of the initial or final "states" (both made by an electron and a proton). The ball players in Figure 26 could be playing in the dark, not seeing the ball, and yet knowing that somehow they had "interacted."

Neutrinos have no electrical charge and cannot scatter off protons in the same way as electrons do. The Electroweak Theory predicted that the **neutral current** process[54] $\nu p \to \nu p$ could also happen, via the exchange of a then hypothetical particle: The Z^0. As soon as neutral currents were discovered at CERN—basing their decision just on this indirect evidence for the existence of the Z^0—the Nobel Foundation rushed to give their somewhat appreciated prize to GW&S. They took an unprecedented risk. The most clearcut prediction of the Electroweak Theory was that the $W^{\pm}$ and Z^0 could actually be produced and directly observed. The Nobel Committee did not make a mistake, the *W*s and the *Z* were later produced and—much more literally—"seen" at CERN.[55] Yet another ingredient of the Electroweak Theory, the Higgs boson, was also discovered at CERN, it deserves its own chapter.

Once one knows that the weak and electromagnetic interactions can be unified into the Electroweak Theory, what could be more tempting than unifying the latter with QCD? With characteristic modesty, physicists call such an endeavor *Grand Unification*, even if it does not include gravity—the even grander enterprise is called a TOE, or theory of everything, it attracts the greatest minds and the worst crackpots with comparably irresistible strength.

The simplest and most convincing *Grand Unified Theory* (GUT) is due to Howard Georgi and Glashow. It unifies quarks and leptons, and explains the relation between their "commensurate" charges (before, the charge of a quark could have been 2/3 and that of the electron 1.36, or anything else). I recall another eminent colleague of mine telling me that, as

[54] So called to distinguish it, for instance, from $\nu_e n \to ep$, in which the lepton and nuclear charges change.

[55] For an entertaining recount of these developments see *Nobel Dreams: Power, Deceit, and the Ultimate Experiment*, by Gary Taubes. Random House, 1987.

he was reading the "preprint" of this work, his heart was pounding, he was awestruck with emotion... until he reached the prediction that "diamonds are not forever"; that is, the proton is unstable, all ordinary matter should one day be gone.

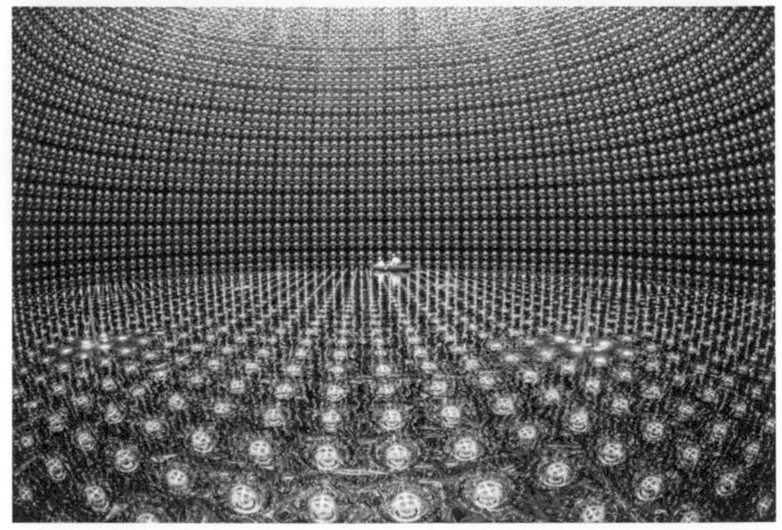

Figure 51 The pioneering IMB detector in the United States [top, ©IMB] and the Kamiokande detector in Japan [bottom, ©Kamioka Observatory]. The Japanese detector, like its photo, was bigger, better . . . , and luckier. It employed the technologies developed for IMB by Larry Sulak and collaborators.

Contrary to my colleague, I thought that *proton decay* was a most beautiful prediction. The expected proton lifetime was many orders of magnitude larger than the age of the Universe. But the Universe contains quite a few protons, whose (quantum) lifetime is the average of a distribution. A tiny fraction of protons ought to decay within an observable time interval, provided one looked at an enormous amount of protons, waiting for a few to decay. Experimentalists were thrilled and constructed enormous detectors, two of which are shown in Figure 51.

Serendipity

IMB and Kamiokande were giant ultra-pure-water pools, protected against cosmic radiation by being built deep underground. Their surfaces were covered by *photomultipliers*, akin to the sensors of a gigantic

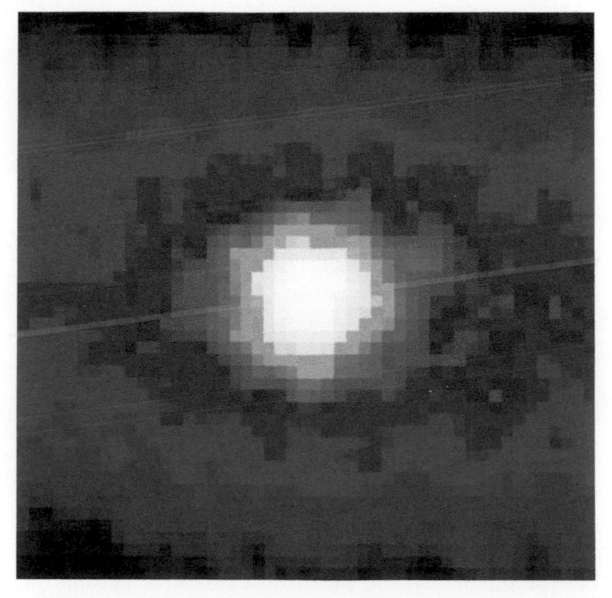

Figure 52 Perhaps the best "photograph" of the twentieth century. It is rather a low-resolution *neutrino-graph* of the core of the Sun seen from underground on Earth by observing, day and night, the neutrinos that the Sun emits (at night the solar neutrinos cross a fraction of the Earth before being observed in the lab). Image obtained by Super-Kamiokande, ©Kamioka Observatory.

photographic camera and sensitive to a variety of light-emitting particle processes taking place in their bulk.

Proton decays were not observed at the level predicted by the mentioned and extremely alluring original GUT, proving it wrong. Either the Goddess was asleep the day She ought to have read the corresponding article, or She does not, after all, exist. On the other hand, somewhat serendipitously, the proton-decay detectors enormously contributed to the advancement of science. They happened to record the neutrinos emitted by the supernova SN1987A, inaugurating neutrino astronomy; "another way to look at the heavens." This kind of detector also discovered *oscillations* of neutrinos made by cosmic rays[56] in the atmosphere and traveling to the detector from above or from below (through the Earth). They also proved right the theory of how the Sun works, by

Figure 53 Red set: A farmer, his pig and a truffle. Blue set: A theorist, an observer (or experimentalist) and the theorist's ideas (or possible discoveries). By the way, you correctly guessed that I am a theorist.

[56] Neutrino oscillations are transformations of one neutrino species into another; for instance, $\nu_\mu \rightarrow \nu_e$. Cosmic rays, for which no convincing theory exists more than a century after their discovery, are mainly high-energy protons impinging on Earth from outer space. Some of their interactions with atmospheric nuclei produce neutrinos, mainly ν_μ and $\bar{\nu}_\mu$.

observing the neutrinos that it makes. I cannot resist the temptation of showing Figure 52, the Sun's core "seen" with neutrinos. The core's radius is about a fifth of the solar radius.

In particle physics, discoveries—serendipitous or not—are generally made by *experimentalists,* in astrophysics and cosmology by *observers*. In both cases there are also the *theorists.* High time to explain the distinctions. This is done in Figure 53. The question is what the similarities and differences between the two sets are. One set consists of a farmer, his pig, and the truffles, the other of the theorist, the experimentalist (or the observer), and the discoveries. The farmer takes his pig to the woods. The pig sniffs around and discovers a truffle. The farmer hits the pig with his bat and takes the truffle away. These are the similarities. The difference is that the theorist scarcely ever directs the experimentalist to woods where there are truffles.

Gauge Theories ★★

The Electroweak Theory, QED (the "electro" part of "Electroweak") and QCD are gauge theories. Here "gauge" means measure, and—by now you will not be surprised—the word refers to something that cannot be measured. The simplest of these theories is QED. In it, the field describing electrons cannot be completely specified, it has a "gauge freedom." Admittedly, this freedom is a truly weird concept (when people talk about freedom, they hardly ever mean what they say). Independently, at each point in ordinary space-time, the electron field is free to sit at any point in an extra "closed" dimension composed of all points along a circle.[57]

The *Gauge Principle* consists of *imposing* the afore mentioned gauge freedom, an act with the most extraordinary consequences. In QED it implies that photons must be massless and that they interact with electrons in precisely the observed way. Another consequence is the conservation of charge and the current carrying it, a theorem originally due to Emmy Noether, one of the few women of science having the fame that they deserve.

In QCD the gauge principle is slightly more complicated: For quark fields, the "circle" of electrons is an ever so slightly more complicated

[57] Physicists, used to the extremely useful formulation of quantum mechanics that employs complex numbers, prefer to refer to this gauge freedom as a choice of "phase" of the electron's field, the φ in $e^{i\varphi}$, for the erudite reader.

space. The principle also implies that gluons are massless, that they couple to quarks or other gluons just as they do in the real world, and that color is conserved. The *W* and the *Z* are not massless. The *gauge symmetry* of the Electroweak Theory must be "broken," as we shall see when discussing the Higgs boson.

The Gauge Principle dangerously sounds like a magic trick. *Things may point in different directions in an unobservable space... and that is why Nature is the way it is.* But it is only a generalization of something one may be used to from much simpler physics. An electrical potential (a voltage) is not directly observable. Only potential *differences* are. Most people recall the difference between introducing a metallic object in a power outlet and then two, one in each hole. Ouch!

Einstein's theory of gravitation (general relativity) is also a gauge theory. But we do not understand it yet at the quantum level. Maybe some "true" extra space dimensions, like the ones invoked in string theories, need to be invoked.

15

A Parenthesis: Is Basic Science Useful?

By now you may—or may not—be convinced that understanding the Universe is interesting. But is it also useful? A not atypical answer is illustrated in Figure 54. This answer is demonstrably wrong.

Many things we take for granted had their origin in fundamental physics. Among so many other examples: The radio, TV, weather and GPS satellites, electrical generators and motors, electronic circuits and computers, cellphones, lasers, X-rays, PET (Positron Electron Tomography), NMR (Nuclear Magnetic Resonance), proton therapy . . . even the universal language of the web: *http* (HyperText Transfer Protocol).

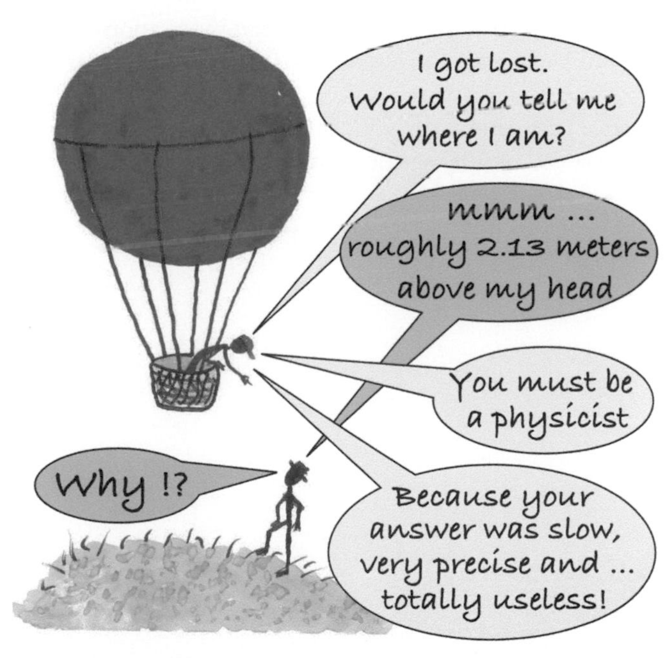

Figure 54 A dialog with an incorrect conclusion.

Enjoy Our Universe: You Have No Other Choice. Alvaro De Rújula.

DOI: 10.1093/oso/ 9780198817802.001.0001

The delay between a theoretical discovery and its practical application varies. From the publication in Newton's *Principia* of his laws of gravitation to their first "application"—the launch of the Sputnik satellite—271 years elapsed. From Maxwell's laws of electromagnetism to Hertz's functioning radio, twenty four years. From the understanding of nuclear fission by Otto Hahn to Enrico Fermi's first nuclear reactor, a mere three years. In a comparably brief interval the http protocol—invented at CERN by Tim Berners-Lee—made the Internet a global means of communication, with an impact only comparable to that of the printing press.

Admittedly, not all fundamental discoveries lead only to user-friendly applications. The oldest example may be the wheel: It is used both in babies' carts and armored vehicles.

Concerning the utility and applications of basic science, a dialog in Plato's *The Republic* (360 BCE) comes to mind. Socrates and Glaucon discuss topics to be taught at a university, after having agreed on arithmetics and geometry:

Socrates: Suppose we make astronomy the third—what do you say?

Glaucon: I am strongly inclined to it, the observation of the seasons and of months and years is as essential to the general as it is to the farmer or sailor.

Socrates: I am amused at your fear of the world, which makes you guard against the appearance of insisting upon useless studies.

16

Back to the Twins

> The distinction between the past, present and future is only a stubbornly persistent illusion.
>
> ALBERT EINSTEIN

Recall that a muon, μ, is an electrically charged particle, basically identical to an electron, e, but for the fact that the muon's mass is $\sim$207 times larger than the electron's. Muons are unstable, they *decay* into electrons and a couple of neutrinos, $\mu \to e\, \nu_\mu\, \bar{\nu}_e$. Muons at rest decay with an *average lifetime* $\tau \simeq 2.2 \times 10^{-6}$ seconds.

Muons are excellent "clocks." Not only are they all identical to each other but, being elementary particles, they have no parts that may malfunction. Neither at rest, nor moving, do muons (after they are artificially made) decay at a fixed time. Their number decreases *exponentially* with time: If it takes a time T for half of the muons to have decayed, half of the remaining muons will decay in the next interval of duration T. And so on and on, until you run out of muons.[58]

Muons freely moving at a constant velocity v along a straight trajectory are observed to decay with a longer average lifetime than muons at rest, $\tau' = \gamma\, \tau$, with γ their Lorentz factor. The punchline is going to be that muons moving at a constant velocity. v, but bent in a magnetic field so that they travel around a curved path (a varying $\vec{v}$ of constant length v), have their lives prolonged in precisely the same way as muons traveling in a straight line: $\tau' = \gamma\, \tau$. All as in the caption of Figure 11.

The first experiment to observe muons decaying as they moved along a closed trajectory was performed at CERN in 1961. They went around and around in a race-track trajectory: A long straight section, followed by half a circular one, another straight section, and the closing half circle. When the experiment was being planned, its design was based on the expectation that the lifetime of muons would be stretched by

[58] This is a probabilistic statement. For an ensemble of N remaining muons it has a statistical uncertainty $1/\sqrt{N}$. Opinion polls quote a similar uncertainty. Sometimes.

Enjoy Our Universe: You Have No Other Choice. Alvaro De Rújula.

DOI: 10.1093/oso/9780198817802.001.0001

the same factor γ along the linear and curved pieces of the racetrack. At the time that was a daring *hypothesis*, in the sense that it had not been observationally checked and there were still reputed scientists who doubted it.

The experimentalists' bet paid off: The experiment worked as planned, the predicted and observed muon trajectories coincided, no muons unexpectedly crashed out of the racetrack on the semi-circular bends and they decayed at the same pace throughout their flight.

Notice that the previous paragraph hid a tacit hypothesis: The *centripetal acceleration* (induced by a magnetic field) required for the muons to negotiate the bends should not modify the expectation that they would have the same velocity-dependent lifetime as straight-moving muons.

The simplest theoretical argument on the possible effect of acceleration on the muons is the one a particle theorist would give. The acceleration is induced by the magnetic field. In the first year of a particle theory course, one learns how to theoretically compute the lifetime of muons at rest. Good students get the observed result. It is not much harder to compute the lifetime of muons moving while embedded in a magnetic field. The result is that, for magnetic fields of the intensity used in the experiments, their effect on the muon's lifetime is totally negligible. If acceleration (as opposed to just velocity) played a role in the life(time) of moving muons, why is the *originator* of the acceleration ineffectual?

The most recent "paradoxical muon" experiment at CERN was performed with $\gamma = E_\mu / m_\mu = 29.327$, with E_μ and m_μ the energy and mass of the muons.[59] It measured the predicted time dilation with a precision of about one part in a thousand. The centripetal acceleration of the muons along their trajectory—a circle this time—was $\sim 10^{18}\, g$, with g the gravitational acceleration on the Earth's surface[60], $g \sim 9.8\, m/s^2$. This apparently enormous acceleration is (by inhuman standards) tiny: It would be many orders of magnitude too small to significantly deform, for instance, an atomic nucleus. Astronauts and Formula 1 racers daringly confront accelerations of a mere few g.

[59] This γ corresponds to a velocity equal to 99.942% of the speed of light. Journalists often use velocities as opposed to γ-values when trying to impress their readers.

[60] Velocity is distance over time and so are the units in which it is measured, e.g., kilometers per hour. Accelerations are changes of velocity with time. Thus, their units are distance over time squared.

Why the Twin Paradox is so paradoxical ★

As anticipated in Chapter 5, twins who move apart from each other and then meet again are "paradoxical" in the sense that both of them may expect to be the one who aged more slowly.

The twins' conundrum is also called the *Clock Paradox*. Google "clock paradox" (quote/unquote, not to add the independent results for clocks and paradoxes) and you get, in 0.38 seconds, "about 713.000 results." The overwhelming majority of them are not the sort of discussion I meant, they rather reflect the *information paradox*, which is a real one: The more is not the merrier.

The above Google experiment substantiates another point: Understanding science is not always so simple. This is my excuse for being so extensive about the twins; they are an excellent example of the struggle between science and the reality it intends to describe. But, should you be a wee bit fed up with the Einsteinian twins and be tempted to jump to the much more digestible Chapter 17, let me spill the beans: the twin who always stays in a *single* inertial reference system is the one who ages faster.

There is an obvious distinction between the twins: The one traveling in a circular trajectory is subject to the *centripetal acceleration* continuously pulling him toward the circle's center. Otherwise he would not have come back to his sister. A very tempting (false) conclusion: It is the *acceleration* of the moving twin and not his *velocity relative to his sister* that slowed down his clock. Recall that velocity effects are "relative," acceleration effects are "absolute." A strong point in favor of acceleration versus velocity.

A *thought experiment* argument against the need to invoke acceleration was proposed by Lord Halsbury. Let us first go back to one of the twins of Figure 11 moving in a straight line from "here" to "there" in the reference frame where the other twin is at rest (as opposed to following a closed path to be back to meet his aged twin). After going from here to there at a velocity $\vec{v}$, he abruptly stops and restarts backward to return from there to here at a velocity $-\vec{v}$. To do this he had to decelerate and reaccelerate back. Were these accelerations what made him find his sister so much more aged? Clearly not, says Lord Halsbury.

Substitute the twins for triplets. As In Figure 55a, triplet T1 is at rest and triplet T2 passes by T1 at a constant velocity $\vec{v}$; they start their stopwatches then. Triplet T3 is far away, moving relative to T1 at a

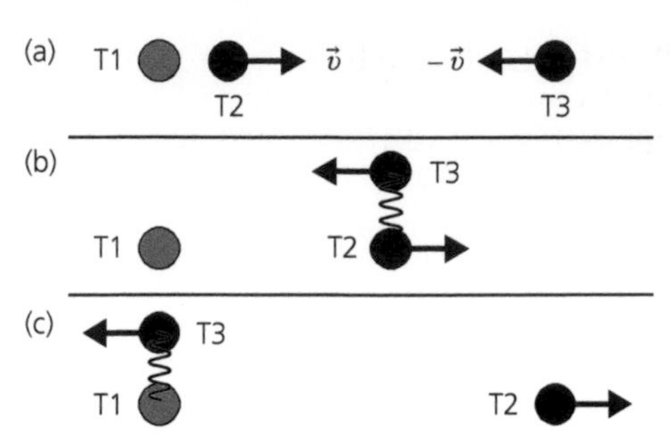

Figure 55 Lord Halsbury's thought experiment with traveling triplets. The wavy lines represent the radio communications between them.

velocity $-\vec{v}$. Later on, T2 and T3 pass by each other, as in Figure 55b. Triplet T2 reads his clock at that moment and notes the result: t'_{T2}. In a time and moving distance both much shorter than the total times and distances involved in Halsbury's "experiment," T2 sends a radio signal to T3 who starts his stopwatch from zero and writes down T2's result, t'_{T2}. Later on, as in Figure 55c, T3 reaches T1, looks at his reading, t'_{T3}, and tells T1 the total result: $t' = t'_{T2} + t'_{T3}$. Triplet T1 checks that $t' = t/\gamma$, with t the time elapsed in his watch during the entire exercise.

This result is exclusively based on Special Relativity. It does not even contain the word "acceleration." Notice that Halsbury uses three siblings, and results as measured by three distinct inertial (unaccelerated) observers. In discussing muons following a circle we had to generalize from three to infinity (the number of infinitesimal steps along a full circle). We are getting tantalizingly close to the crux of the matter.

The twin or triplet who stays put throughout in *only one* inertial frame is the one who ages more rapidly.[61] The other guy either suffers an acceleration or uses Halsbury's trick to transmit information from an inertial frame (T2's) to another one (T3's).

To conclude: When they meet at the end of the tail, why is one twin older than the other and not (illogically) also the other way around? Because theirs was not a real paradox, since its formulation contained a MacGuffin, a subtle (?) asymmetry: *Only one* twin is always in a *single* inertial frame. There is a consequent asymmetry in the result: It is the other twin who ages more slowly.

[61] More generally, of a collection of arbitrarily moving clocks, the one that runs fastest is the one whose *mean squared velocity* (the average of v^2 along its trajectory, measured relative to a given single inertial frame) is the smallest.

Presumably the simplest resolution of the Paradox was found as late as 1957 by Charles Galton Darwin, a grandson of Darwin the evolutionist. It is the obvious solution: If two people do not understand each others' point of view, let them talk it over. The twins cannot talk when they are far from each other. But they can send a light signal to the other guy on their respective birthdays. Light certainly catches up with either twin. From the knowledge of their relative velocity and the time of arrival of the signals, each twin can figure out how the other one is aging. Neither during their journey nor at the end of it do they face any problem in reaching the conclusion that one ages faster than the other.[62]

An interesting consequence of the resolution(s) of the apparent paradox is that accelerometers are not needed to decide whether a certain motion is uniform or accelerated. Comparing clocks suffices.

Not everybody is satisfied with the fact that the resolution of the Twin Paradox requires *only* Special Relativity. This is Einstein's "fault." He very significantly thickened the plot.

Twins according to Einstein

Back to the lady twin at rest and the other one flying on a straight line out from "here" to "there" and then back. When they meet again, she is older than him. Get ready for a good surprise. Suppose that the gentleman twin, at rest in his "reference system," has placed clocks at rest in it. These clocks, compared with the one she carries with her, measure her getting *younger* than him on the way out, and also on the way in. Trouble. But, lo and behold! His clocks see her aging suddenly by an awful lot at the time (in her reference system) when she takes the abrupt turn from moving out to moving in. And her total aging, including the sudden jump, is such that when she is back she is older than him, by the "right" amount.[63]

Einstein published his opinion on the twins only after his 1916 work on General Relativity, when he figured out the brilliant, though

[62] The exercise, which I do not do here, requires solving pairs of linear equations and/or drawing space-time diagrams. It is even more convincing if the birthday messages are sent with an agreed light-frequency, whose blue- or red shifts provide a double check.

[63] The above is not a total absurdity, it is a consequence of the "relativity of simultaneity": Two events occurring simultaneously in one reference system are not seen happening at the same time in a different system. By having one twin jump systems we are looking for trouble. Sudden aging is not (in this case) a biological reality.

somewhat unnecessary, resolution of the quandary encountered in the previous paragraph.

Recall that Einstein postulated that acceleration and gravity are equivalent. Though he did not specify how—probably thinking that it was too obvious—he computed the time dilation due to a "pseudo-gravitational" effect (equivalent to the acceleration) felt by the lady twin who is measured (according to the records kept by the other one's clocks, distributed throughout his reference system) to age suddenly. The answer is precisely the amount of extra time required to recover the fact that one specific twin, at the end of the tale, is older than the other!

Why did Einstein use the expression *pseudo*-gravitational? Presumably because he knew the effect was "real" only in the mind of the twin trying to understand what the matter was with the other one. Yet, in spite of the "pseudo," Einstein's result was a most remarkable test of the consistency and complementarity of special and general relativity.

Back to experiments, doable or not

The muon-decay experiments provided results relevant to the Twin Paradox, but only as a byproduct. They were intended to measure the *magnetic moment* of the muon, whose meaning (for the analogous case of the electron) is discussed at length in Chapter 13. Similarly, an experiment by Robert Pound and Glen Rebka, intended to measure the effect of gravity on photons ("gamma rays" in this case), also gave a "twin byproduct." General Relativity predicts the amount by which the frequency of a photon increases as it "falls" (and the other way around).

Pound and Rebka observed with exquisite precision the gamma rays emitted or absorbed by the nuclei of the Fe^{57} isotope of iron. They either "dropped" them from the top floor of their building at Harvard to a detector in the basement—which increases their frequency, or vice versa—which decreases it. Their results agreed with the prediction of General Relativity.

Pound and Rebka also measured how their results varied with **temperature** (the Fe atoms are in a crystal lattice and vibrate more intensely the higher the temperature). At room temperature the vibrating atoms undergo accelerations of order $10^{16}\,g$. The accelerations were measured to affect the results by less 10% in comparison with relative velocities. Subsequent experiments improved this limit to 0.01%. This gave the final punch to possible effects of acceleration on the rate of clocks. Relative velocities once again won the context.

Newton worried about water filling a bucket. If you rotate the bucket around its axis, the water spills. It also does if you move the bucket with a sufficient acceleration. But relative to what did you rotate (or otherwise accelerate) the bucket and its water? Relative to the ensemble of all matter in the Universe, say some. Take it all away and the water will never spill. Not an easy experiment to perform.

The "cheapest" solution to the Twin Paradox reintroduces a universal rest system—distinct from the observationally excluded good old ether—in which one but not the other twin is "absolutely" at rest, eliminating the temptation of asking why it is one twin and not the other who ages slowly. I do not know how to test these ideas. If there is no way, they are not convincingly relevant to physics.

Respectful of the opinions of others—hundreds of thousands of them, according to Google—I let the readers draw their own conclusions on the Paradox, particularly if they have an astronaut twin.

The algebra behind the relativity of time ★★★

Witness its three stars; This section is quite algebra-heavy. It is only intended for readers not hostile to a little math.

Back to Figure 5 we may reconsider why the difference between fathoms and miles makes sense, even if they are both measures of the same but more abstract—or deeper—entity: Distance. The two units obviously make separate sense, at least on a ship, to the captain and the fisherman. But if we moved both of these people to empty space, there would be no water surface: The distinction between "horizontal" direction, x, and the perpendicular (or orthogonal) one, the depth y, would disappear. The distance d between two objects, though, would still make "absolute" sense. This means that, in empty space, we could change x and y by rotating the orthogonal directions we call "horizontal" and "down" and, yet, always see the same d. That is $d^2 = x^2 + y^2$ stays put, even if x and y change when "rotating" (in the rotated coordinates, denoted as primed, $x'^2 + y'^2$ is also d^2). Trivial thoughts on geometry!

Special relativity also has a geometrical aspect. An event is something that happens at a specific point in space and at a given time. Let the distance in space between two events be d and the difference of their timings be t. Define the *interval* between the events as Δ, with

$\Delta^2 = t^2 - d^2$ in natural units. This is analogous to the definition of distance in the previous paragraph, *but for the minus sign*. Next, imagine different unaccelerated observers moving at different velocities and looking at the two events. Their measures of distance and time-intervals would be different, but their observed values of Δ^2 are the same, so that the two observers see light moving at the same speed.[64] This fact of Nature has fascinating consequences. Back to the twins one last time.

Suppose my twin has a ruler and a clock. On the ruler she has two marks, one for "here" (where she is), one for "there" (at the opposite end of the ruler, at a distance d from the "here" mark), as in Figure 11. I am passing by and she sees me go from here to there at a constant velocity, in a time that she measures to be t. The squared interval between the two events (I pass by here, I get there) is $\Delta^2 = t^2 - d^2$, which she can rewrite as $t = \sqrt{\Delta^2 + d^2}$. The distance "from me to me" is $x' = 0$, I am where I am. Thus, the time it takes me to go from one to the other end of my sister's ruler, as measured by my watch, is $t' = \Delta$, corresponding to my not having moved (recall that Δ is the same for both siblings). And now for the inevitable conclusion: Since $t' = \Delta$ is smaller than $t = \sqrt{\Delta^2 + d^2}$, my elapsed time is smaller than my sister's: I am younger than her at the end of my voyage.

To be even more explicit. My time t' and my displacement, $x' = 0$, are such that $\Delta^2 = t'^2 - x'^2 = t'^2$. For my sister, in a time t, I have moved at a velocity v a distance $x = vt$ and $\Delta^2 = t^2 - x^2 = t^2 - v^2 t^2 = (1 - v^2)\, t^2$. Since Δ ought to be the same for both of us, $t'^2 = (1 - v^2)\, t^2$, that is $t' = t/\gamma$, with $\gamma = 1/\sqrt{1 - v^2}$, the Lorentz factor.

As anticipated in footnote 19, the energy, E, and momentum, p, of a particle of mass m are related like times and distances are. The energy ($E = m\gamma$) versus velocity plot in Figure 9 is a way of stating that $m^2 = E^2 - p^2$, with m (unlike E and p) independent of the relative velocity between the particle and the observer.

The word *relativity* refers to time, space, energy, and momentum being "relative"; that is, different for observers moving in different ways. *Absoluticity* would have been a horrible but better name, for the true "surprise" of relativity is that certain observables, such as mass and the speed of light… are "absolute," not relative. The absolute nature of a specific *velocity*, that of light in vacuum, seems to be particularly

[64] Indeed, re-introducing c for clarity, an observer would measure $d = ct$ and another $d' = ct'$. The values of $\Delta^2 = c^2 t^2 - d^2$ and $\Delta^2 = c^2 t'^2 - d'^2$ are equal (and vanish).

astonishing. But only at first, then one gets used to it and finally one concludes that it could hardly be otherwise. Indeed, only relative velocities are meaningful, for there is no way to measure one's speed relative to *nothing*. If no laws of Nature contradict that, including the ones describing light, then all of Special Relativity follows, including the "absoluticity" of c.

After this strenuous algebra we deserve a rest. The next few chapters are all downhill.

17

Some Instruments of Macro-Physics

Humans, birds, seals, and even dung-beetles navigate by the stars. The latter do it even at a fixed angle of their choice, relative to the axis of the Milky Way . But, on our planet, only humans are capable of much more than the cited adorable creatures.

From time immemorial, humans have catalogued stars, constellations, planets, and comets, speculated on their influence on their lives and, more scientifically, predicted the motions of celestial objects, including the prognosis of lunar and solar eclipses. One of the many early examples of the results of such endeavours is the decoration of the ceiling of the tomb of the Egyptian pharaon Senenmut, painted almost three and a half millennia ago, see Figure 56.

Admirably sophisticated astronomical instruments have existed for centuries. At the left of Figure 57 we see the remains of Ulugh Beg's device to spot the passage times and positions of celestial objects. This was done through an opening at the top of a marble railing, with a sliding eyepiece used to determine the elevation above the horizon. The observatory is in Samarkand (Uzbekistan) and dates from the 1420s. Predictably, it was destroyed by religious fanatics in 1449. Its remains were rediscovered only in 1908.

At the right of Figure 57 is Galileo's telescope—an instrument that he did not invent, but which he did improve. It is kept in a museum in Florence that is now named after him. Galileo's telescope was not destroyed, but brought him well-known "problems" with the Inquisition. With this instrument he observed the phases of Venus and the moons of Jupiter, decisive blows to the officially accepted *geo*-centric theory of the "solar" system (*solar* as in *helio*-centric, the correct theory). The last traces of official opposition to heliocentrism by the Roman church disappeared in 1835 when Galileo's works were finally dropped from the *Index* (of books forbidden by the Vatican).

Enjoy Our Universe: You Have No Other Choice. Alvaro De Rújula.

DOI: 10.1093/oso/ 9780198817802.001.0001

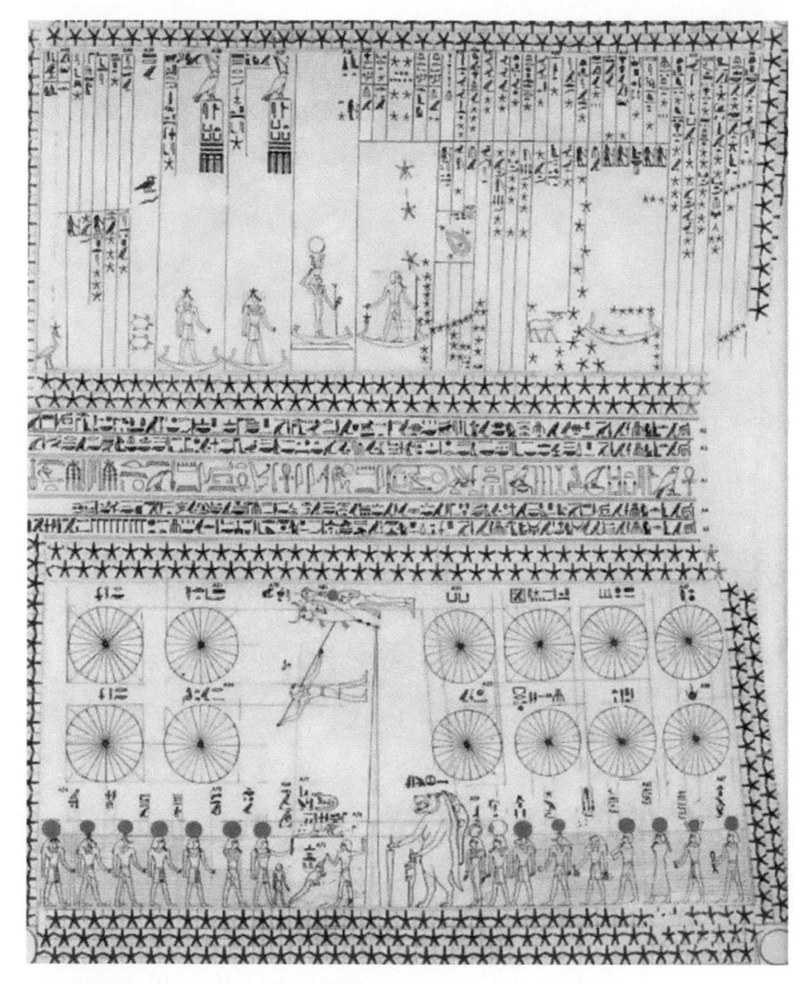

Figure 56 An astronomical "chart" in the tomb of Senenmut. Recognizable (by experts) are Sirius, Orion, the Big Dipper, Jupiter, Saturn, Mercury, and Venus. Mars sits on an empty boat, perhaps because of its then retrograde motion (all planets rotate around the Sun in the same direction, like the two needles of a clock. But from one planet others *appear* at certain times to move "backward"). The months of the inundation, planting and harvesting seasons are represented as various circles. https://commons.wikimedia.org/wiki/File:Senenmut.jpg.

Galileo's Telescopes
The cracked lens is mounted in centre

Figure 57 The remains of Ulugh Beg's sextant, and Galileo's telescope. https://commons.wikimedia.org/wiki/File:Ulugh Beg%27s Astronomic Observatory.jpg; *Studies in the History and Method of Science*, Volume II. Edited by Charles Singer (Oxford: Oxford University Press, 1921).

Microwave antennas and satellites

Not all "telescopes" use photons in the visible-light range to explore the sky. We have seen examples in the first part of Chapter 14 and Figure 51 for which the "messengers" were neutrinos. Back to photons as the observing tool, photons in the microwave (MW) range of frequencies are not visible by eye. The microwave antenna shown in Figure 58 was used by Erno Penzias and Robert W. Wilson in their 1964 discovery of the Cosmic Background Radiation (CBR), also called the MWBR. Puzzled by a signal that they thought was a mysterious noise, they are seen in the photograph intensely scrutinizing their antenna. The reason is that they first thought the "noise" was induced by what they politely called "a dielectric material," deposited by invasive New Jersey pigeons. Instead of four-letter slime, what befell them was the 1978 Nobel Prize. A good

Figure 58 The Bell Labs' antenna with which the microwave background radiation was discovered by Penzias and Wilson. They are shown in the picture pondering what the origin was of what they saw. Antenna ©NASA. https://pixabay.com/en/dove-flying-peace-olive-branch-41260/.

example of serendipity, since what their antenna was originally built for was the detection of radio waves reflected by Echo balloon satellites.

The next big discovery concerning the CMB was made in 1992 with the COBE satellite: The fact that the radiation is not entirely isotropic, a subject that we shall discuss in great detail. This satellite and the first anisotropies it observed are shown in Figure 59. In this figure, unlike in Figure 13, the entire sky is projected onto an image bound by an ellipse.

Gravitational-wave interferometers

One consequence of Einstein's theory of General Relativity is that the equations of Figure 1 imply the existence of gravitational waves, emitted by any accelerated object. These waves can be interpreted as deformations of space-time that, in a vacuum, travel at the speed of light. Interestingly, Einstein spent his life oscillating between "believing" or not in this prediction of his theory. The decisive argument, that the

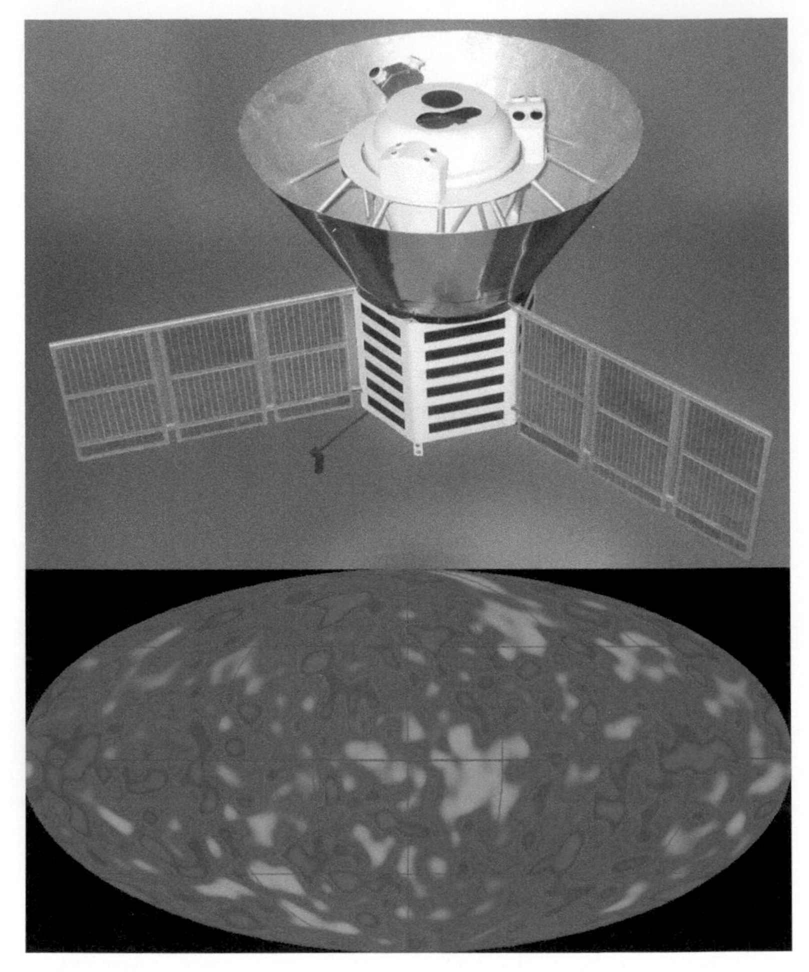

Figure 59 NASA's Cosmic Background Explorer and the anisotropies it observed in the cosmic background radiation. ©NASA.

waves could deposit energy in a "sticky bead," was made by Richard Feynman in 1957. Even in the periods in which Einstein believed in "his" waves, he thought that they would be too weak to ever detect.

Two objects of recent fame in this respect are the LIGO Gravitational-wave interferometers of Figure 60, resurrected from their initial technical and economic limitations by Barry Barish who, as this book was in press, got the 2017 Physics Nobel Prize, along with Rainer Weiss and

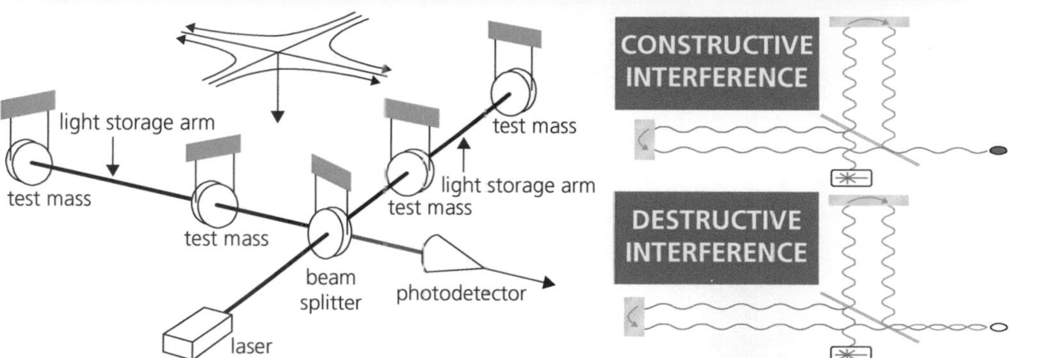

Figure 60 Top: The LIGO gravitational-wave detectors in a high desert in Washington State and in swampy Louisiana. Bottom: a schematic view of an antenna and its working principle (https://commons.wikimedia.org/wiki/File:Gravitational wave observatory principle.svg, drawing by Cmglee).

Kip S. Thorn. These antennae are very sophisticated objects exquisitely "suspended" to avoid "noise" from Earth tremors, both natural—continuous diminutive quakes—and "unnatural"—passing trucks, people firing guns... A super-powerful laser fires pulses of infrared light along vacuum tubes. The light is split by a semitransparent mirror to travel along two 4 km-long arms. "Test mass" mirrors reflect the light back to the beam splitter, which also works as a beam merger, "normally" making the two outgoing light waves cancel each other (they add precisely out of phase) so that no signal reaches the photodetector downstream.

If and when a gravitational wave happens to reach us from outer space, it distorts space-time in such a way that the light waves merge successively in and out of phase, generating an oscillating signal in the photodetector. Believe it or not, the device can detect relative variations of the 4 km length of the antennae's arms as small as 1 part in 10^{21}. This corresponds to measuring a length of 10^{-16} cm, one hundred millionth of the size of an atom, or the size of a proton divided by 1000. No wonder Einstein did not believe any of this could ever be done!

18

The Discovery of Gravitational Wave Emission

Conventional stars are made of ordinary matter, most of which is ionized: The atoms, at high temperature, have lost some or all of their electrons, which roam by themselves. A *neutron star* is mainly made of neutrons, held together—like ordinary stars—by its own gravity. These stars are much denser than, say, the Sun. Many of them have a mass about 40% larger than that of our star, but a radius of a few kilometers. As a result, a liter of average neutron-star stuff has a mass of roughly 4×10^8 ... tons!

A large fraction of stars are in *binary systems* of two of them orbiting each other. Some of the most interesting binaries are those made of two-

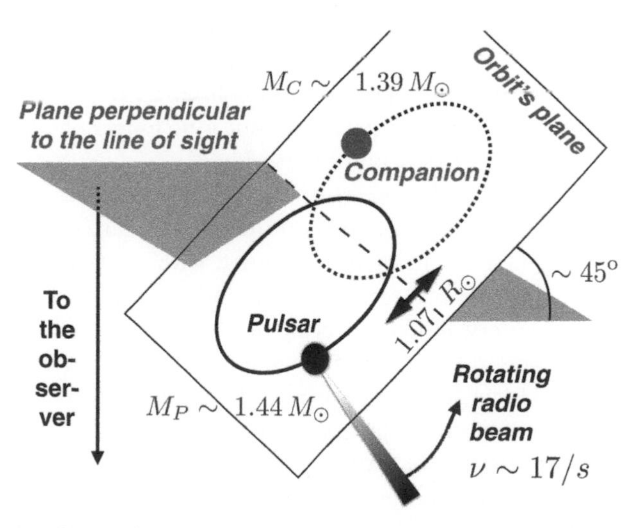

Figure 61 The Husle–Taylor binary. Two neutron stars, one of which was observed as a pulsar, from an angle of ~45° relative to the orbit's plane.

Enjoy Our Universe: You Have No Other Choice. Alvaro De Rújula.

DOI: 10.1093/oso/9780198817802.001.0001

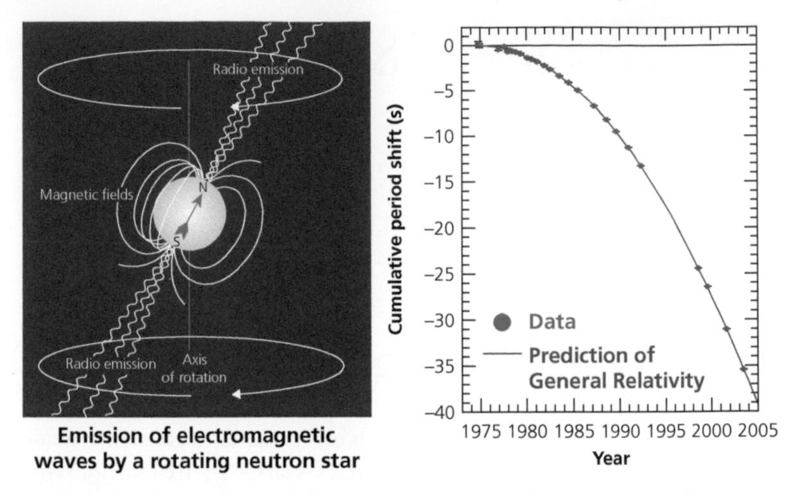

Figure 62 Left: Emission of electromagnetic waves by a magnetized, rotating neutron star, ©http://www.nobelprize.org. Right: Shortening of the orbital period of PS 1913+16 (totaling ~36 s in ~29 y), due to the emission of gravitational waves. https://commons.wikimedia.org/wiki/File:PSR B1913%2B16 period shift graph.svg, by Inductiveload.

neutron stars. The first neutron-star binary to be discovered, illustrated in Figure 61, is called PS1913+16 (the numbers are sky coordinates, it was found in 1974, not in 1929). The PS stands for "pulsating star," or pulsar; one of the the two stars of PS1913+16 *was* a pulsar. Why "was"? A pulsar is like a beacon. This particular one is no longer observable: Its radiation does not currently point toward us.

The way a pulsar pulsates, which is not understood in gory fully satisfactory detail, is illustrated in Figure 62. Like the Earth, the neutron star has a magnetic field not quite aligned with its axis of rotation. This rotating field emits electromagnetic radiation that sweeps us in successive time intervals, like the beam of a lighthouse.

Unlike the one in Figure 58, the large-dish microwave antenna in Arecibo, Puerto Rico, does not require an illustration. The reason is that "everybody" has seen it in a James Bond film (*GoldenEye*). Russel Hulse was a student of Joe Taylor, hunting for pulsars from Arecibo. Pulsars produce radio signals that go "beeep-beeep-beeep." Hunting them is routine. But one lucky day Hulse found a pulsar that, every few hours, went to and fro from "beeeep-beeeep-beeeep" to "beep-beep-beep." And yes, he soon realized that they had discovered the first binary pulsar.

The Hulse–Taylor (HT) binary pulsar consists of two neutron stars, one of which is a pulsar. The illustration in Figure 61 shows that the orbits are very elongated ellipses. The orbital velocities vary over a large range, from 110 to 450 km/s (relative to the center of mass). Like a train's horn, the beeps of the pulsar appear to arrive with a higher frequency when it is moving toward the observer, and at smaller frequency when it moves away. Thus the tell-tale varying beep-rate observed by Hulse.

Many binary pulsars are impressively extreme in comparison with other star or planetary systems. The HT pulsar rotates in 59 nanoseconds, the duration of its "day". The distance between the stars varies from 1.1 to 4.8 solar *radii*, as they complete an orbit in 7.75 hours. These extreme characteristics imply that, by carefully analyzing the record of pulsar's beeps, one can measure with flabbergasting precision not only the orbital properties previously referred to, but also the individual star masses (approximately 1.4 times the mass of the Sun) and various "non Newtonian" observables only predictable and understood on the basis of General Relativity.

The HT pulsar system provides an exquisite test of Einstein's theory. Not only that, the large accelerations of the stars in their orbits imply that the system loses energy by the emission of gravitational waves.[65] This shrinks the orbits and shortens the period of revolution by a predictable amount. The prediction and the observations are shown on the right of Figure 62. They are in perfect agreement. Einstein's theory—this time a prediction that he did not always believe—is brilliantly vindicated!

In some 300 million years the orbits of the HT binary will have shrank so much that the stars will coalesce. We may not be around to see the resulting gravitational fireworks, but just wait until the next chapter.

[65] For the discovery and analysis of the first binary pulsar, Hulse and Taylor received the 1993 Nobel Prize. So did Antony Hewish in 1974 for the discovery of pulsars, made by his thesis student Jocelyn Bell Burnell. Dame Jocelyn was invited to the 1993 Nobel celebrations. By Joe Taylor, I presume. Unfairly tiny recognition of her pioneering work.

19

The Direct Detection of Gravitational Waves

What could be "bigger and better" than the coalescence of two neutron stars? The coalescence of two unexpectedly massive stellar black holes! The gravitational waves made by such an event were observed for the first time by the LIGO antennae on September 14, 2015, see Figure 63. Given the date of the observation, this "event" is called GW150914.

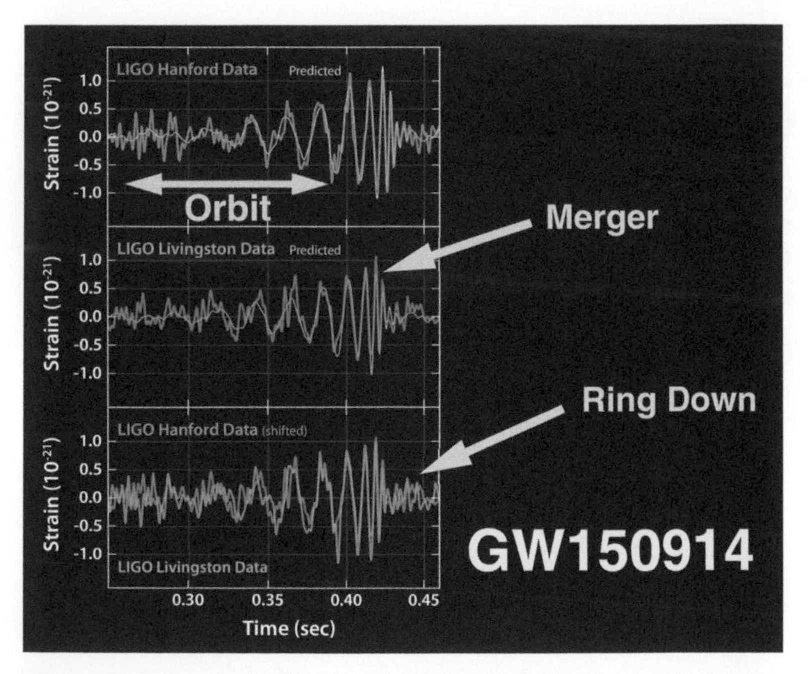

Figure 63 Top and middle: The first observed coalescence of black holes, recorded by the LIGO gravitational antennae. Bottom: superposed signals, corrected for the different times of arrival, at the velocity of light, to the antennae (6.9 milliseconds) and for their different orientation.

Enjoy Our Universe: You Have No Other Choice. Alvaro De Rújula.

DOI: 10.1093/oso/9780198817802.001.0001

Figure 64 Light (photons, γ), neutrinos (ν), and gravitational waves ($g_{\mu\nu}$); Our three ways to "look at the stars." Windows from http://maxpixel.freegreatpicture.com/Water-Scene-Blue-Sea-Window-Ocean-Seen-View-164630.

Like two neutron stars, two black holes orbiting around each other emit gravitational waves in a "chirp": A rising frequency and increasing intensity as the objets get closer and move faster. In the LIGO data of Figure 63 the chirp is seen at the end of the "orbit" period, about a tenth of a second of a celestial dance having lasted millions or billions of years. The "merger" or coalescence of the two black holes into a single one lasts even less time. Finally, the newly born merged black hole vibrates for a while, still emitting gravitational waves: A "ring down." The entire recorded process lasts ~0.2 seconds.

The properties of the holes can be extracted from the GW150914 data to a precision of about 15%. The masses of the parent holes are $29\,M_\odot$ and $36\,M_\odot$ (recall that $M_\odot$ denotes the mass of the Sun). The merged hole has a mass of 62 suns. No typos here. Indeed, three solar masses are "lacking." Not really: An energy of $3\,M_\odot\,c^2$ was emitted in the form of gravitational waves. During a fifth of a second, the (gravitational) "luminosity" of this black hole merger was 50 times the (light's) luminosity of *all* the stars in the visible Universe. A perfect storm in the framework of space-time!

The previous theoretical understanding of gravitational-wave signals and the analysis of this particular one are also monumental feats—on which we have no space to dwell. After light and neutrinos, gravitational waves have opened *a third way to look at the sky*, as in Figure 64. With no fresh political interferences, the future of gravitational-wave astronomy should be much much "brighter than a thousand suns."

As I witnessed it, the progress in our understanding of gravity appeared to proceed in fast-forward motion. The reason is that many of my professors at the university where I studied (Madrid) were not

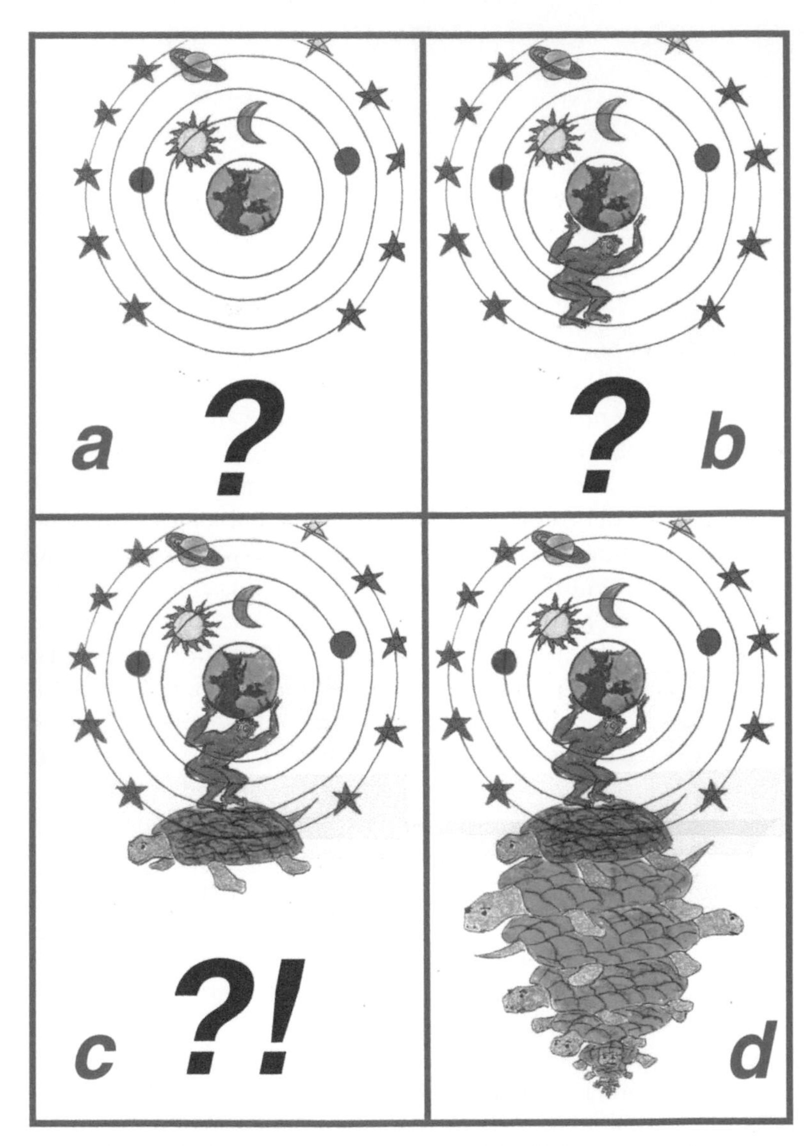

Figure 65 Ancient views on gravity and the Universe around us.

entirely up to date. A conversation between an inquisitive student and the professor teaching AA3 (the third course on Advanced Astronomy) could well have been the following:

Professor: The Earth is in the middle, surrounded by successive spheres on whose surfaces the Sun and the Moon, the inner planets such as Venus, the outer planets such as Saturn, and finally the stars… turn around us. All as in Figure 65a.

Student: But, Professor; what stops the Earth from falling down?

Professor: I taught that in AA1. The Earth stands on the shoulders of a giant, Atlas. See Figure 65b.

Student: And how can Atlas hold up there?

Professor: Imbecile, that I explained in AA2. Atlas stands on an enormous turtle, as in Figure 65c.

Student: But…

Professor: Don't even ask. I'll give you the answer straight away: It is turtles all the way! As in Figure 65d.

References

[1] M. Dacke, E. Baird, M. Byrne, C. H. Scholtz, & E. J. Warrant, *Current Biology*; Volume 23, Issue 4, February 18, 2013, page 298

20

Some Instruments of Micro-Physics

Telescopes and microscopes are used to observe things much larger or much smaller than we are. Microscopes have a limitation that telescopes do not have: As their resolving power increases, so does the likelihood that they destroy what they are trying to observe. In a microscope the light shone into the observed object is reflected and seen or photographed. An ocean wave may be scattered back by a large rock but would be unaffected by a thin pole planted on its way: There is no reflected wave constituting a "picture" of the pole. Similarly, to see the details of some object you have to shine on it light with a wavelength smaller than the object, so that its structure may be "resolved." A short wavelength means a high frequency, and a high frequency implies a high energy, as we recalled at the beginning of Chapter 9.

Objects smaller than atoms, such as their nuclei, can only be well resolved by shining on them light whose constituent photons have such a high energy that they generally affect the object, exciting or breaking it. You can try "shining" the observed object with something other than photons: Neutrinos, electrons, protons, and what not. The results are similar.[66] In high-energy (or elementary-particle) physics a typical observation ends up with the object under study entirely blown up to smithereens.

The instruments going beyond microscopes in studying the structure of objects smaller than atoms or the properties of objects of vanishing dimensions (elementary particles) are *accelerators* and *colliders*. Both machines would be more aptly called "energizers." Accelerators send whatever they accelerate to high energy onto a *fixed target* containing the object under study. Colliders consist of two opposite-direction accelerators making identical or different particles ... collide.

[66] In $c=1$ units the energy and momentum of a photon are equal $E=p$. For a massive particle $E=\sqrt{p^2+m^2}$ and the photons' Einstenian energy-frequency relation, $E=h\,\nu$, is substituted by $p=h\,\nu$.

Enjoy Our Universe: You Have No Other Choice. Alvaro De Rújula.

DOI: 10.1093/oso/9780198817802.001.0001

Some accelerators (including the pairs of which some colliders are made) are *linear* and boost particles, once, along a straight line. Others accelerate particles again and again as they travel around along a spiral or circular path.

Fixed targets are typically solid or liquid. They have a density much bigger than that of the beams in an accelerator or collider. Thus, in studying collisions of similar particles, the rate at which beam-beam collisions take place in colliders is much smaller than the rate in the collisions of a beam with a fixed target, much the same as in firing the pellets of a shotgun against a wall, or against the oppositely-moving pellets of a second shotgun. This is the good news for the use of fixed-target machines as opposed to colliders. The bad news is that the energy of the particles in a collider is entirely "used" in the collisions, while in a fixed-target case it is partially "wasted" in the collective recoil kinetic energy of the beam and target fragments. Consequently, colliders are the winners of the highest-energy context.

The first working circular accelerator was made by Ernest Lawrence, who patented it in 1934. It is shown, in his hand, in Figure 66. Since then, accelerators have made considerable progress and grown in size, energy, and the intensity of their beams. The accelerator complex at CERN is shown in Figure 67. Its largest component is the LHC (Large Hadron Collider), a circular machine 27 kilometers (17 miles) in circumference. The protons (or lead nuclei) that it collides are successively pre-accelerated in a series of smaller accelerators, one of which (the proton-synchrotron, PS) has been operational for longer than half a century.

CERN straddles the countryside of France and Switzerland. Some of its facilities, the LHC in particular, could not possibly be built overground, mainly because the terrain is not flat; building an up-and-down meandering accelerator would be difficult. So the LHC is located in an underground tunnel built—in the usual money-saving fashion—to host a previous collider, see Figure 68. As a consequence, one does not see the LHC from the air, as in Figure 69, where the Geneva airport is visible but the SPS and LHC rings are added by (electronic) hand.

Also overlaid on the photo of Figure 69 is a dotted line separating France (on top) from Switzerland. From a distance the two countries seem similar, but a closer view would reveal that the Swiss farms are clean and flowery, while many of the French ones are littered with

Figure 66 Lawrence with his first proton accelerator and the device itself. His was a "cyclotron" wherein the accelerating particles follow a spiral trajectory before they are "ejected." Courtesy of Berkeley Lab.

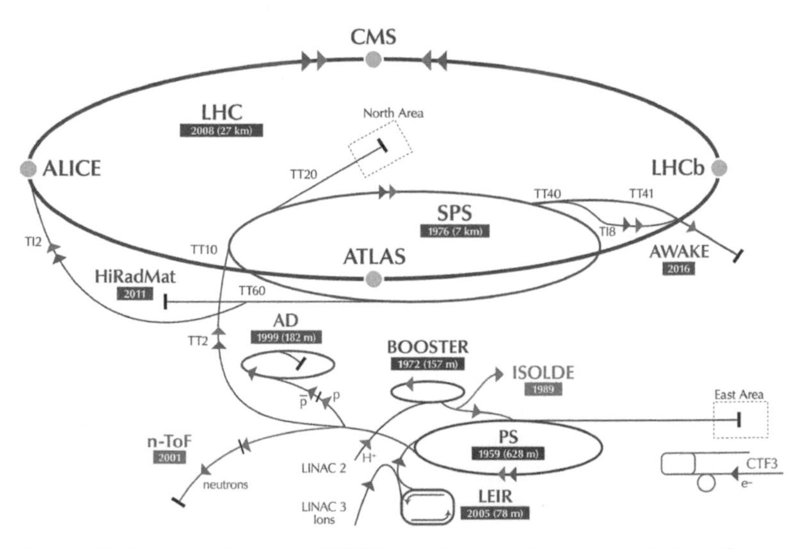

Figure 67 The complex web of CERN accelerators. Protons are "injected" into the two LHC rings after having been pre-accelerated in a linear accelerator (LINAC 2), and three circular ones: BOOSTER, PS, and SPS. Many ion accelerators and fixed target facilities are also shown. The AD is an antiproton *decelerator*, aimed at precise studies of antimatter. Courtesy of CERN.

Figure 68 A worker checking the digging of the tunnel in which the LHC is. Given the monster drilling machine, the guy was no doubt a bull-fighter. Courtesy of CERN.

Figure 69 An areal view of the location of the SPS and LHC underground tunnels (the smaller and larger circles) and the airport of Geneva. The dotted line separates France (northeast and above) from Switzerland. Courtesy of CERN.

rotting cars and the like. Concerning the wines produced here and there, however, the situation is rather different, more equilibrated.

A close view at Figure 69 reveals that an LHC proton would cross the French-Swiss border six times as it completes a turn. Traveling at an energy of some 7000 times its rest energy, its velocity is very close to that of light[67] and it makes a full turn some 11,000 times per second. Thank goodness there are no custom officials to stop progress (or tourists) here.

A microscope and a magnifying lens have the same light-bending operating principle, which need not be recalled. That may not be case for an accelerator. The LHC, for instance, has two types of main components, repeated over and over along its circumference. The first of these are *accelerating cavities*, one of which is shown in Figure 70. The lower part of the figure illustrates the way it works. Protons travel in

Figure 70 An LHC accelerating cavity being tested. The blue plate changes "polarity" from negative to positive as a proton (whose charge is positive) goes through its central hole. Two energizing kicks per passage. Courtesy of CERN.

[67] That is $\gamma = 1/\sqrt{1 - v^2/c^2} = 7000$, making v only 10^{-6} of a percent below c.

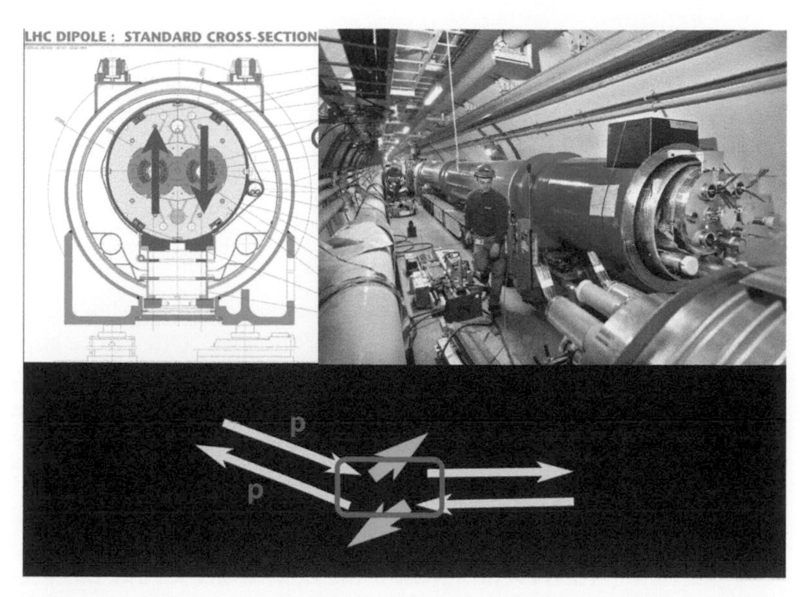

Figure 71 A 17 meter-long (blue) LHC magnet being installed in the LHC tunnel. A diagram of its cross section shows the opposite magnetic field directions in the two beam pipes. Particles of the same charge moving in opposite directions are bent in the same centripetal direction. Courtesy of CERN.

an accelerator inside vacuum *beam pipes*. A cavity contains a negatively charged plate with a hole to let the arriving protons through. The plate attracts the positively charged protons, accelerating them to higher energy. As soon as a *bunch* of protons goes through, the plate is charged positively. This repels the protons and gives them a second accelerating kick from behind.

The second main component of an accelerator is a large set of magnets. Some are used to "focus" the beams: Reducing their transverse dimensions. Others, called *dipoles* steer the two oppositely moving beams of the LHC along their circular trajectories. One of them is shown in Figure 71. Embedded in the magnet are the two beam pipes. In one of them the protons are subject to a downward-directed magnetic field, in the other beam pipe the field points upward. The consequence of this arrangement (opposite velocities, opposite field directions) is that both beams are steered so that they follow the accelerator's circumference.

21

The LHC and its Detectors

There are four very large and a handful of smaller detectors located deep underground at the four collision points of the LHC. One of the large cavities that house them is shown, during its construction, in Figure 72. It is connected to the surface by a wide vertical shaft, through which pieces of the detectors are lowered to be assembled underground.

Shown in Figure 73 is one of the large detectors, CMS, during its build-up. The LHC detectors are akin to a sophisticated camera, taking "pictures" of collision "events." Tens or hundreds of particles are typically created in each event. *Created* is the right word, for the particles

Figure 72 The underground cavity that now houses the Atlas experiment at the LHC, during its construction. Courtesy of CERN/Atlas.

Enjoy Our Universe: You Have No Other Choice. Alvaro De Rújula.

DOI: 10.1093/oso/9780198817802.001.0001

Figure 73 The CMS detector being assembled underground. Courtesy of CERN/CMS.

are newly made in the collision, they are not pre-existing pieces of the colliding particles. Contrariwise, most of the pieces (all but some easy-to-make photons) in a collision between, say, two cars, were there before: Wheels, carburetors, and what not.[68]

Electrically charged particles describe curved trajectories in a detector's magnetic field. The curvature determines the sign of their charge and their momentum. Their velocity can also be measured from their "time of flight" between detector elements at accurately known distances. Combined with their momentum, this determines their mass and, thus, their identity.[69] Photons and electrons deposit their energy in "electromagnetic calorimeters." Neutral hadrons do it in "hadronic calorimeters." Calorimeters measure the particles' energies. All these very many different detectors are distributed in an onion-like structure, with an "inner tracker" closest to the narrow beam-pipe wherein the accelerated particles collide.

[68] To create a particle of nonzero mass m, a minimum energy $m\,c^2$ is required. A car collision has enough energy to produce zillions of massive particles, but it is distributed between the cars' very many elementary constituents, no two of which collide hard enough to produce anything but some (massless) photons, e.g., in sparks.

[69] In $c = 1$ units $p = m\,v/\sqrt{1-v^2}$, so that $m = (p/v)\sqrt{1-v^2}$.

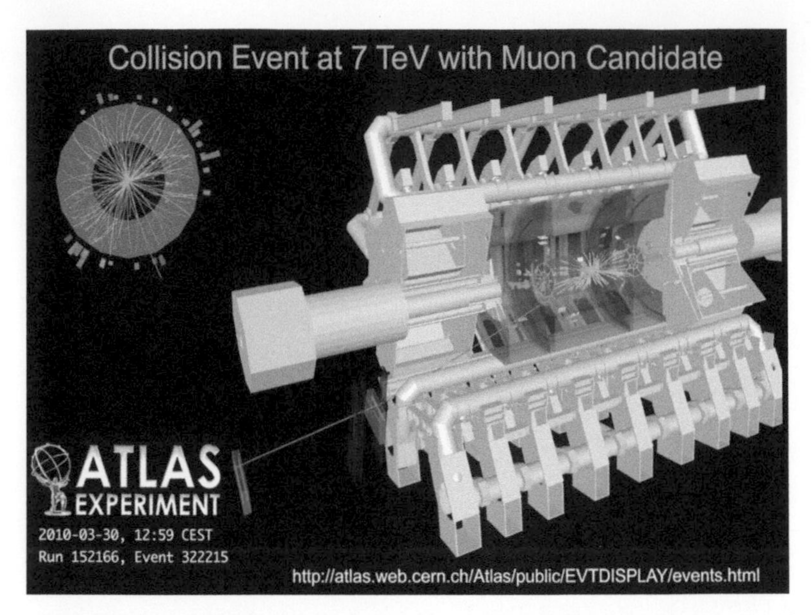

Figure 74 An early event in Atlas, showing tracks of charged particles, one of them a very penetrating muon. The circle on the upper left shows the event projected onto a plane perpendicular to the colliding beams. There, the yellow rectangles reflect energies deposited in a calorimeter. Courtesy of CERN/Atlas.

The construction of an accelerator such as the LHC, and of its detectors, is an enterprise requiring tens of years and thousands of people. In this sense, it is comparable to the building of cathedrals, to mention only European temples of old. A main difference is that—physicists being more modest than architects—cathedrals are built on ground, while large high-energy-physics (HEP) facilities are assembled underground, as shown in Figure 75. A pioneering precedent is *La Cathédrale Engloutie* (the sunken cathedral), a piano piece by Claude Debussy.

One of the first Atlas events is shown in Figure 74, where we see an image of the detector superimposed with the event's computerized reconstruction. It contains a very penetrating charged particle—a muon—that reaches the outer "muon detector," shown only as the two "hits" in it (the green slabs).

After decades of collaborative design and construction of the LHC and its detectors, the first collision event in each experiment appeared in the computer screens of the "control rooms" (on November 24,

Figure 75 A couple of high-energy physicists trying to see their detector/-cathedral. Top: from a photograph by Sander van der Wel in https://commons.wikimedia.org/wiki/File:Bury your head in the sand.jpg. Bottom: Photo by Luis Miguel Bugallo Sánchez in https://commons.wikimedia.org/wiki/File:2010-Catedral de Santiago de Compostela-Galicia (Spain) 2.jpg.

2009, at the modest "injection energy" of 450 GeV). An example of the consequent reaction by some of the experimentalists is shown in Figure 76. At the lower right corner of the figure we see the two people who had just taken an entire night's shift.

Besides being extreme examples of technical prowess (skill) and prowess (courage), the accelerators and detectors we have discussed are "a beauty." Not surprisingly, they inspire artists. An example is shown in Figure 77, showing the LHC complex and CMS as seen by Sergio Cittolin, a physicist a bit influenced by Leonardo, as admitted in the figure.

Figure 76 Joy explodes as physicists see a first collisional event at the LHC. Notice the two people who had done all of the previous night's work. Courtesy of CERN.

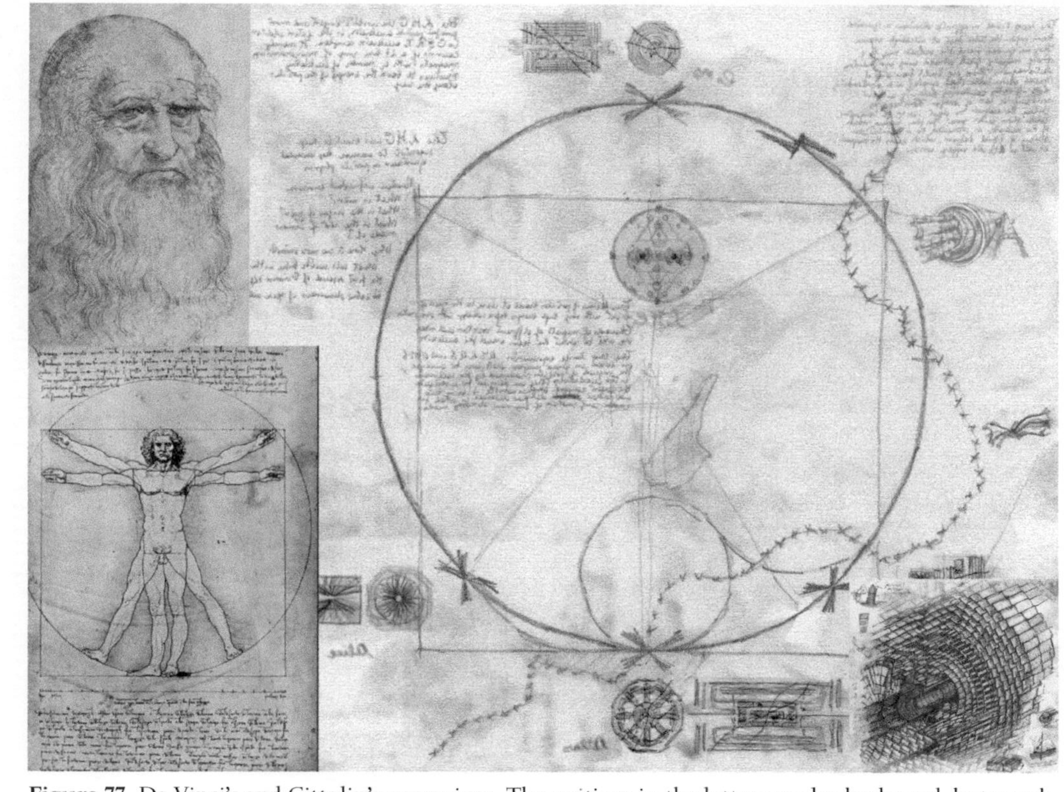

Figure 77 Da Vinci's and Cittolin's engravings. The writings in the latter are also backward, but much more humorous. Courtesy of S. Cittolin.

22

The “Higgs” Boson, and its Field in the “Vacuum”

The Electroweak Theory of the weak and electromagnetic interactions is an example of a construct whose basic symmetry is *spontaneously broken.* What this means is that the equations describing the theory are symmetric, but their “solutions” are not. An example of this phenomenon is given in Figure 78. A ball at the bottom of a glass with the shape shown in the figure’s left side would stay where it is. The ball could oscillate up and down in the same way in any direction; that is the symmetry (all directions are equivalent and/or the cup remains the same if rotated around the vertical axis). The lowest energy state corresponds to the ball at rest sitting at the origin of the plane defined by the black arrows in the figure. This lowest energy state is rotationally symmetric.

On the right side of Figure 78 there is another possibility, also with the same rotational symmetry; the lower part of a conventionally shaped wine bottle. But this time, a ball sitting at the central cusp would be in an unstable situation. It would tend to fall “spontaneously” in one direction or another. Once it has fallen the symmetry is broken. The ball sits in a particular point in the plane defined by the black arrows. The state of lowest energy still corresponds to the ball at rest, but is no longer symmetric. The “theory” is rotationally symmetric, but its lowest-energy solution has a “preferred” direction where the ball sits.

In the electroweak theory the “bottle” describes the **Higgs field** φ and how it interacts with itself. In its lowest-energy state, also called *the vacuum*, the Higgs field has a non vanishing value, denoted φ_0.

The Mother of all Substances

We have just seen that, in its lowest-energy stable state, the Higgs field of the Electroweak Theory—and of this Universe of ours—has a non-zero value, φ_0, called its *vacuum expectation value.* Concerning the room in which

Enjoy Our Universe: You Have No Other Choice. Alvaro De Rújula.

DOI: 10.1093/oso/9780198817802.001.0001

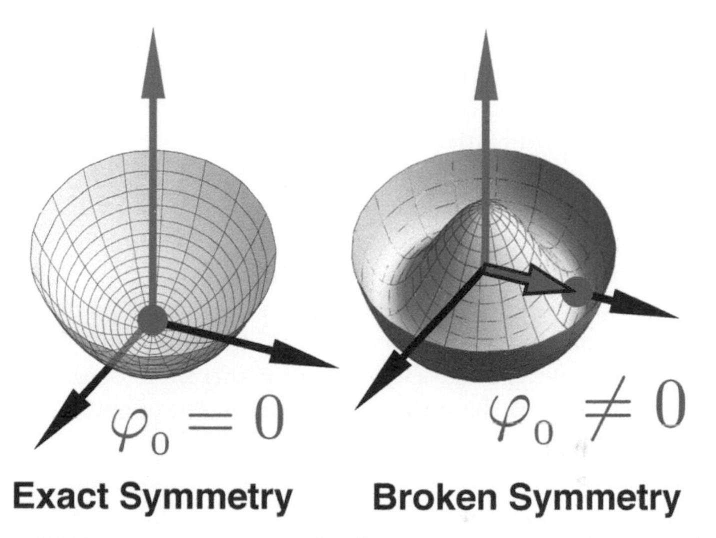

Figure 78 Spontaneous symmetry breakdown; the Higgs mechanism. Left: A symmetric case with a symmetric state of lowest energy. Right: Another symmetric situation, but with a state of lowest energy that breaks the symmetry: Where the ball sits defines a specific direction.

you are presumably sitting and as stated at the beginning of the second part of Chapter 7: The room is permeated by something that you cannot smell, hear, taste, touch nor see, but of which you are definitely aware. It is the Earth's gravitational field. You could get rid of this annoying field by moving your room to a place in the extragalactic realm where the gravitational field is extremely negligible.

In the mentioned "middle of nowhere," you have disposed of the gravitational field, but *not* of the Higgs field's φ_0. This means that the vacuum is not empty! **The vacuum is a substance.** We shall see that the vacuum, naively the simplest of all things, is not truly understood yet. But we know quite a bit about how the Higgs field acts.

Allegedly, the Higgs vacuum expectation value, φ_0, uniformly permeates the Universe in outer space as well as anywhere else, including inside you. And the masses of the elementary particles that have mass—with the exception of the Higgs boson itself and perhaps neutrinos—are proportional to φ_0. If φ_0 vanished, the masses of intermediate vector bosons, quarks, and charged leptons would also vanish. Much as photons "couple" to particles in proportion to their charges, the Higgs field (and its value in the vacuum) couples to these particles in proportion to their masses. A particle interacts with the vacuum, permeated by the non-vanishing φ_0, and that interaction generates its mass. This is often compared to a sort of friction, one of many extremely bad analogies.[70] In my opinion the one in Figure 79 is the worst of all.

We have been discussing the Higgs field and its value in the vacuum. All this sounds very peculiar. Is any of it testable? As we saw in the beginning of Chapter 14, a relativistic quantum field has other manifestations, one of them as a particle, in this case the *Higgs boson* and its creation and destruction. The decay (destruction) of a Higgs boson into other (created) particles provides a way to test whether the astonishing things we have been saying about the vacuum are sensible or not.

Figure 79 The winning entry to a British 1993 competition to explain mass generation by a Higgs boson and/or Mrs. Thatcher. It would be difficult to misrepresent better the underlying physics. There is no sense in which inhabitants of the vacuum gather around a massive particle. A CERN cartoon, courtesy of CERN. http://www.hep.ucl.ac.uk/ djm/higgsa.html.

[70] A friction transfers energy from the moving object (such as a bullet) to the retarding one (such as air). The interaction of a particle with the vacuum Higgs field does not.

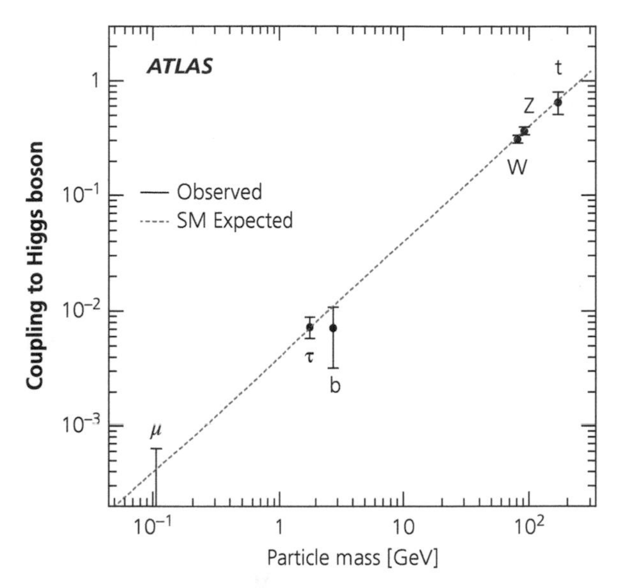

Figure 80 As predicted in the Standard Model (SM), the measured couplings of a Higgs boson to various particles are proportional to their masses. Results of the ATLAS experiment. Courtesy of CERN/Atlas.

The couplings of Higgs bosons to all but the lightest elementary particles have by now been directly measured with considerable precision at the LHC, by observing the probabilities for this particle to decay in different ways: $H \rightarrow \mu^+\mu^-$, $H \rightarrow \tau^+\tau^-$, $H \rightarrow b\bar{b}$, $H \rightarrow W^+W^-$, $H \rightarrow Z^0 Z^0$ and $H \rightarrow t\bar{t}$. As shown in Figure 80, the measured couplings to these different objects are perfectly compatible with being proportional to their masses, as required if the masses are due to an interaction with the vacuum value of the Higgs field. This need not have happened. But for "the Higgs" of the Standard Model, predicted more than half a century ago, it had to happen. And it did!

Making Higgs Bosons

Alas, there are no Higgs boson mines. The predicted average lifetime of a Higgs is 1.6×10^{-22} seconds, too short to measure directly. Like most of the unstable elementary particles, Higgs bosons must be *made* if one happens to be interested in them.[71] Their mass being so large (some 125 GeV, roughly 133 proton masses) they have to be made in very high energy collisions. So far, only the LHC can, on this planet, be a Higgs factory.

The official announcement of the discovery of the boson by the ATLAS and CMS experimental teams at the LHC was made on July 4th 2012, to an enthusiastic crowd that included many of the "fathers" of the boson: François Englert, Gerard Guralnik, Carl Hagen, and Peter Higgs. Tom Kibble could not attend and Robert Brout was no longer alive.[72] There has never been a scientific discovery attracting so much media attention. But scientific announcements generally pale in comparison with, for instance, the launching of a new iPhone, a Super Bowl Final, or a *clásico* (a Barça/Madrid soccer game).

The main production mechanism of Higgs bosons in proton-proton collisions is the complex one depicted in Figure 81. Two gluons within the protons coalesce to produce a $t\bar{t}$ quark pair that fuses into an H. The fact that the coupling of the H to the very massive top quarks is large results in this mechanism being dominant. Two relevant decay modes of the H are also shown in the figure: to two photons and—mediated by fast-decaying Zs—to four charged leptons.

The two "discovery modes" of the Higgs boson were the ones in Figure 81. An actual candidate event of the two-photon channel $H \to \gamma\gamma$ is shown in Figure 82. From the energies and directions of the photons it is possible to reconstruct the mass of the photon-pair, which may or may not be the mass of the H. The name of the game is to extract the "signal" from an enormous "background" of pairs of photons made in many other different ways. Similar considerations apply to other Higgs decay patterns.

How come the discoverers of the Higgs boson were so confident that they had seen "it," to the point of inviting the theorists who had invented it? This is something they do not tend to emphasize. The calculations underlying the mechanisms shown in Figure 81 are highly non-trivial and could not have been made by many an experimentalist

[71] The most notable exceptions are the relatively long-lived muons. Dozens of them traverse you every second. They are the decay products of other particles—mainly pions—made by cosmic rays—mainly protons—colliding with air molecules—mainly oxygen and nitrogen—in the atmosphere—mainly high up there.

[72] Englert and Higgs received the 2013 Nobel Prize for their "mechanism" and consequent boson.

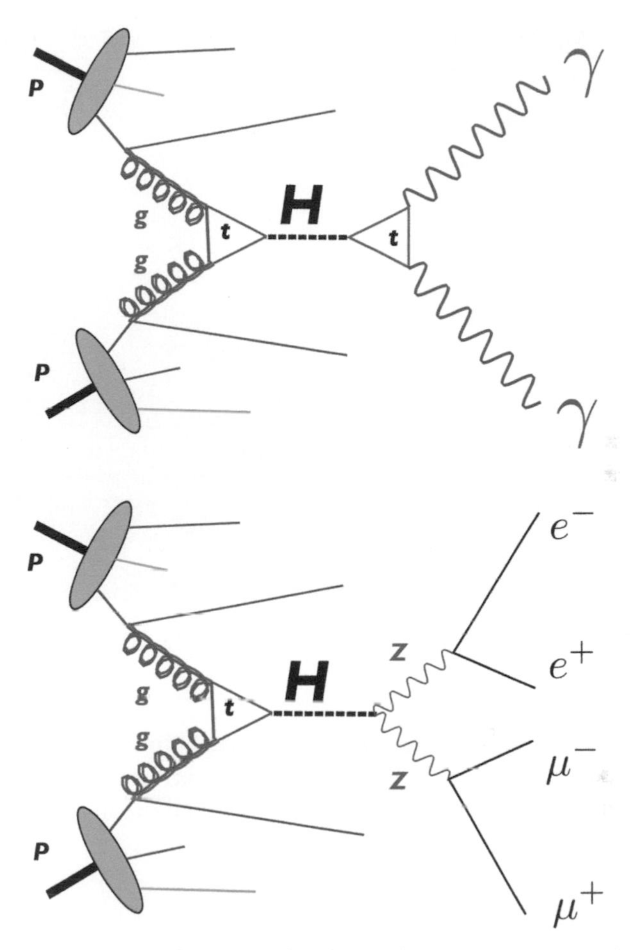

Figure 81 Production of "a Higgs" by gluon "fusion" via a top quark "loop" (a $t\bar{t}$ pair fusing here into an H). Top: decay into two gamma rays. Bottom: decay, via two Zs, into four leptons: A $\mu^+\mu^-$ and an e^+e^- pair.

(that not many theorists could contribute to the construction of the collider or its detectors is even more true!). That the rate at which the bosons were produced agreed with the theory was one decisive reason to trust the interpretation. One of the experiments used the sophisticated "quantum entanglement" of the decay products in the four-lepton decay channel—again worked out by theorists—to boost their findings

Figure 82 Reconstruction of an event with two very energetic photons (γ). Courtesy of CERN/CMS.

above the officially accepted "discovery level." The calculation of the "uninteresting" backgrounds also required extremely laborious and crucial contributions from insufficiently thanked theorists.

23

Today's Standard Models of Particles and Gravity

We are all in the gutter, but some of us are looking the stars.
Oscar Wilde [1854–1900]

With the discovery of the Higgs boson and the direct observation of gravitational waves it would seem that the two Standard Models of particle physics and general relativity are complete. Not true. An extra undiscovered particle—called the **axion**—is required for the particle's model not to predict, contrary to observation, a large violation of the "time-reversal" symmetry of the fundamental laws, which we discussed in Chapter 13. We do not have a convincing quantum theory of gravity. Thus, there are to-dos for current and future physicists, even in the standard realms. And more so outside them: We do not know what **dark matter** is, nor do we understand dark energy at all.

To proceed, we follow Wilde's advice, as Hubble did, see Figure 83.

Enjoy Our Universe: You Have No Other Choice. Alvaro De Rújula.

DOI: 10.1093/oso/9780198817802.001.0001

Figure 83 Edwin Hubble and the Hooker telescope at Mt. Wilson Observatory (California). Left: https://en.wikipedia.org/wiki/File: University of Chicago Basketball Team, Intercollegiate Champions, 1909.jpg. ©The University of Chicago Archives. Right: Photo by Andrew Dunn (http://www.andrewdunnphoto.com/), Creative Commons Attribution-Share Alike 2.0 Generic license.

24

The Expansion of the Universe

We have already glanced at the Universe in Chapter 6 and to its Cosmic Background Radiation in Chapters 6 and 17. The next question regards the extent to which we understand how the entire Universe "works."

As a University student, Edwin Hubble, shown in Figure 83, was a law major and an outstanding football player. But he decided he would be better off as an astronomer (times have changed) and he studied galaxies, then called "nebulae," with the telescope shown in the same figure. He proved that, contrary to previous thinking, these nebulae were ensembles of stars akin to our own galaxy and at large distances outside it, populating what he poetically called *The Realm of the Nebulae*. We would now prosaically call it the visible Universe.

The first observations of the Universe "at work" culminated with the publication by Hubble in 1929 of what was to become known as "Hubble's law," although it was derived by Georges Lemaître a couple of years earlier on the basis of the same observations. Moreover, a fraction of the data Lemaître and Hubble used was obtained by Vesto Slipher in 1917. One of many examples of the phenomenon discussed in Chapter 2: Fame goes to the *last* discoverer.

In its simple original form of Figure 84 Hubble's law relates the distance d from us to a given galaxy and the velocity v at which it *appears* to be moving relative to us: $v = H_0 d$, with H_0 the "Hubble constant." The longer the distance to a galaxy is, the larger its apparent velocity of "recession." This sounds pretty simple, but in astrophysics measurements and their meaning are never this trivial.

Some tempting misconceptions

Suppose that you explode a bomb in empty space. Its pieces rush away in all directions at various speeds. Next, take a picture of these pieces at time t after the explosion. The ones that were ejected with a velocity v

Enjoy Our Universe: You Have No Other Choice. Alvaro De Rújula.

DOI: 10.1093/oso/9780198817802.001.0001

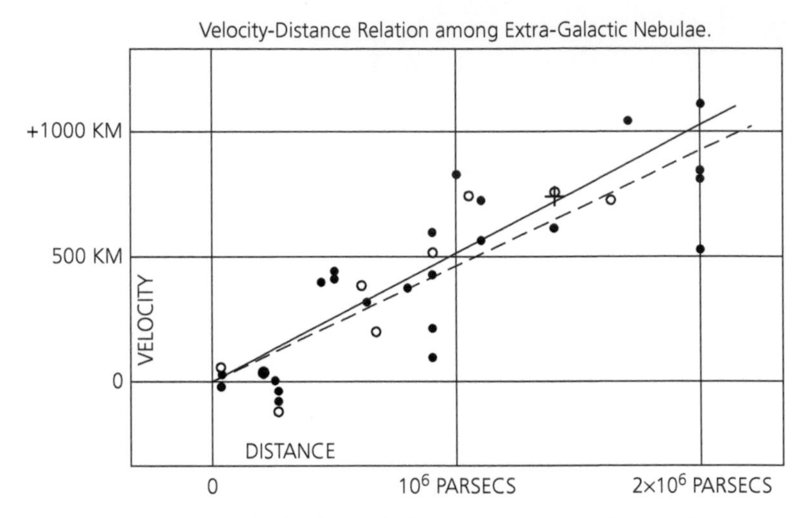

Figure 84 The original plot by Hubble, with apparent velocity of recession versus distance (a parsec is 3.26 light years or 3.086×10^{13} km). E. Hubble, Proc. Natl. Acad. Sci. U.S.A. *15*(3) [1929] pp. 168. Notice the large spread of the data relative to a straight line (a would-be exact Hubble law). Also, some close-by galaxies have $v < 0$, they are approaching us.

are at a distance $d = v\,t$ from the explosion's locus. This is correct for all different values of v: Any of the velocities of "recession" from where the bang occurred. Thus, the various distances to the explosive "center" are proportional to the velocities, or vice versa: $v = d/t$.

Notice that $v = d/t$ looks very much like Hubble's law $v = H_0\,d$, suffice it to identify $H_0 = 1/t$. In the abominable units that astrophysicists favor, the most recently determined value of H_0 is ~ 67.8 km per second per megaparsec. More transparently, $t = 1/H_0 = 14.4$ billion years. In the analogy of galaxies resulting from the explosion of a bomb, this t is the time since the explosion took place. The result is quite close to the currently estimated age of the Universe, $t_0 \sim 13.8 \times 10^9$ years. The zero in t_0 is the peculiar notation used by cosmologists. It means *now* and not $t = 0$, the beginning of time.

All this sounds like A Big Bang Theory of the birth and expansion of the Universe. And it is almost perfectly incorrect. Incidentally, the expression *Big Bang* was satirically introduced in cosmology by Fred

Hoyle, who did not believe in an expanding Universe and who was obviously not unaware of the expression's innuendo in vulgar slang.

Understanding how the Universe works in some detail requires getting used to a few concepts that are not immediately obvious. As a preview, let us counter a few of the errors we may have been making:

- Distant galaxies are receding from us but not in the naive sense of having velocities away from the point where we happen to be. It is the space between galaxies *itself* that is increasing with time.
- In the evolution of the cosmos, atoms and galaxies reached in the past an equilibrium size. It is relative to such fixed dimensions that the space between distant galaxies expands.
- There is no space in which the Universe was born. Space is a feature of the Universe and was born with it. Thus, the Big Bang did not happen in some particular place; Trump's Tower in New York, say.
- There is no time *before* the Big Bang. Time (or more precisely, space-time) is a feature of the Universe and was born with it.[73]

Distances and redshifts in cosmology

The distance d to a galaxy cannot be measured with a meter tape. It is evaluated step by step with the help of a variety of objects, such as some specific stars: Variable ones called Cepheids at the shorter distances, very luminous explosively dying stars (supernovae of Type Ia) in galaxies much further away. The absolute luminosity of these stars (how much light they shine) is fairly well understood. The comparison of this luminosity with the apparent one (how much of their light reaches our telescope) results in an estimate of the *luminosity distance*, d.

The relative velocity between us and a given galaxy cannot be measured in the same sense as that between our car and the road. It is measurable in the same sense as the velocity between our car and a policeman's (Doppler) radar. In both cases, the relevant velocity-like observable is the **redshift**, a measure of the amount by which the

[73] There are non-excluded theories in which the Universe expands and recollapses again and again, with no initial time. It they hypothesize a *First Bang*, a time before it is also a meaningless notion.

wavelengths of emitted and observed radiation differ.[74] The origin of this difference may be a relative velocity between the light-emitter and the observer, the expansion of the Universe, or a combination thereof. It is useful to understand it first in the relative-velocity case.

If we were to listen to a clock that is moving away from us, its tick-tock would appear to be stretched: Since every successive tick or tock is emitted from further away, its arrival to our ear is delayed, see Figure 85. Light of a given color is like a clock, it has a period; the interval of time between successive maxima of its wave. And the period of light emitted by a source moving away from us is also stretched. So is its wavelength, the distance traveled in one period.

Call λ_e the wavelength of an emitted light, generally corresponding to a specific atomic transition in the surface of an observed star. And

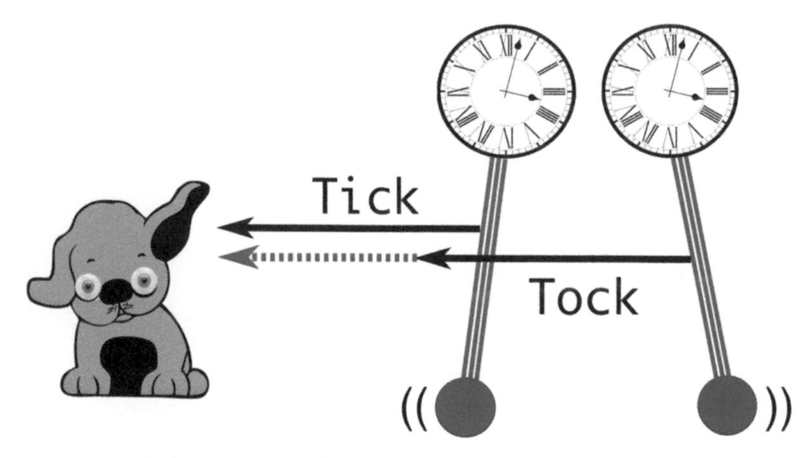

Figure 85 Relative to the listener, a clock is moving to the right. To reach the ear, the sound of a Tick travels a shorter way than the sound of the subsequent Tock. The heard Tick-Tock-Tick time (its period) is stretched. So is the wavelength of the sound (the distance in space between two successive Ticks). All this applies to light waves as well. Listening dog: http://www.publicdomainpictures.net/view-image. php?image=77774&picture=puppy-dog. Clock: https://pixabay.com/en/clock- old-clock-antique-clock-face-1605224/.

[74] Here and elsewhere we are tacitly using the observed fact that the laws of Nature are "universal": The same everywhere. Thus, atoms in a distant place locally emit or absorb light of the same mono-chromatic wavelengths as measured in a lab, here.

let λ_r be the wavelength of the light received by the observer. Explicitly figuring out the "result" of Figure 85, the ratio of observed and emitted wavelengths is $\lambda_r/\lambda_e = 1 + z$, with $z \simeq v/c$ for light waves[75] and $z \ll 1$. The result for sound is the same, with c substituted by its speed.[76]

Of the colors of visible light, red has the longest wavelength. Hence the name "redshift" for the wavelength lengthening of an object moving away from the observer. Notice that in Hubble's Figure 84 the redshifts of some close-by points are negative: These are galaxies moving toward us, their light is actually blueshifted.

The two phenomena illustrated in Figure 86 contribute to observed redshifts. The two upper lines describe the "naive" red and blueshifts we just discussed. In cosmology they are due to the so-called *peculiar velocities* of the observed galaxy and the one we are sitting in. These are velocities

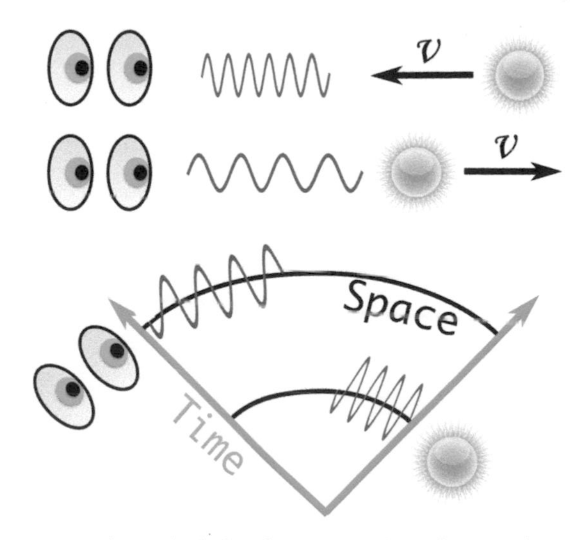

Figure 86 Top two lines: the light of an approaching (receding) object appears to be blueshifted (redshifted). Bottom: The elongation of space in an expanding universe stretches the wavelength of light as it travels. An emitted blue light of wavelength $\lambda_e = 475$ nm may be received as red ($\lambda_r = 650$ nm). The redshift was $z \simeq 650/475 - 1 = 0.5$ and space had fattened by a factor of two between the times of emission and reception!

[75] The correct result for light waves and any z is $z = \sqrt{(1 + v/c)/(1 - v/c)} - 1$.
[76] A difference between sound and light is that a supersonic object traveling toward you may hit you before its sound warns you. There is no superluminal equivalent.

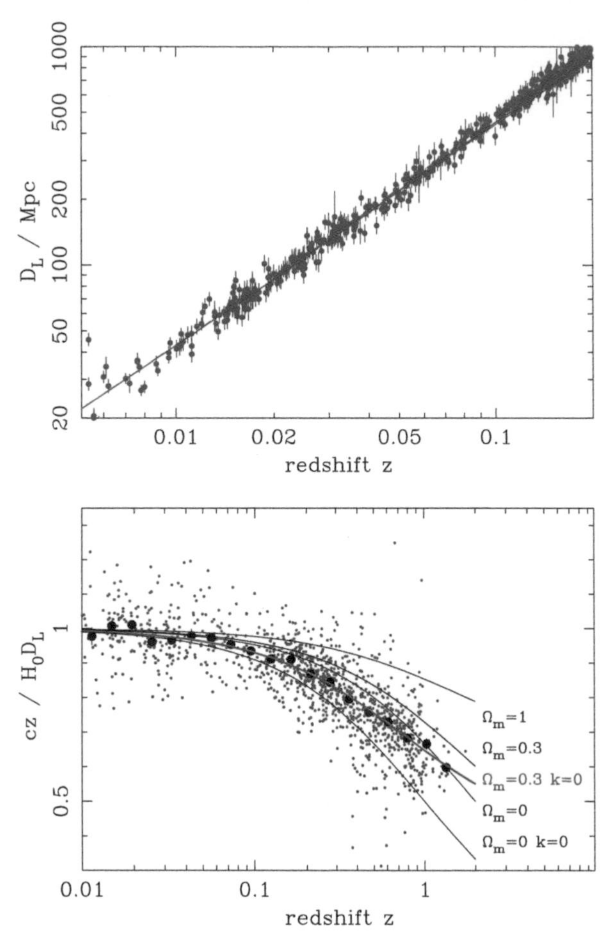

Figure 87 Luminosity distance versus redshift for Type Ia supernovae. Top: The linear Hubble law for $z \ll 1$. Bottom: Deviations from linearity for larger z. Black dots: Observations averaged in z-intervals. Black lines: Various excluded models of the composition of the Universe. A Universe dominated by matter and vacuum energy (the red curve) is favored. Figure 22.1, p. 357, Partigiani et al. Chin. Phys. C40, 100001 (2016). ©2016 Regents of the University of California and ©2016 Chinese Physical Society and the Institute of High Energy Physics of the Chinese Academy of Sciences and IOP Publishing Ltd.

relative to the local reference system in which the Cosmic Background Radiation is isotropic.[77]

The diagram at the bottom of Figure 86 describes the truly fascinating concept; *the cosmological redshift.* The line in space going from the observed star in a distant galaxy to the observer is represented by the black arcs. Space is being stretched by the expansion of the Universe as time (the green arrows) goes by. General Relativity implies that the wavelength of the traveling light is stretched with time in precisely the same way as space itself is. Much as if the wave had been painted on a rubber band that one lengthens by pulling its ends apart.

The peculiar velocities of galaxies are of the order of hundreds of kilometers per second, corresponding to redshifts v/c of order 10^{-3}. Cosmological redshifts of observed supernovae reach $z \sim 2$. Based on them, a modern version of Hubble's observations is shown in Figure 87. The horizontal axes are redshifts, extending to $z = 0.2$ ($z = 2$) in the upper (lower) panel. The vertical scale on the upper panel is the luminosity distance. The figure shows a nearly perfect *linear* Hubble law, distance and redshift are proportional for z between 0.02 and 0.2. At $z < 0.02$ the cosmological redshift is less dominant and the effect of the peculiar velocities (a scatter of the data relative to the red line) is visible. The lower panel shows the deviations from linearity for the larger z values.

The lines in the lower panel of Figure 87 correspond to the predictions of General Relativity with a few choices of the contributions to the energy density of the Universe of matter and the cosmological constant, Λ. All models fail to fit the data except the one whose average density is 30% in (ordinary plus dark) matter and 70% in Λ. We shall comment on all of this in much more detail in the first section of Chapter 27.

[77] This is the absolute *rest system* at each location in the Universe, a concept to which we shall return. Not a contradiction with relativity in empty space: The Universe is not empty. It is filled, among other things, by the CBR.

25

Finding Cosmic Fossils

The distance versus redshift measurements imply that the Universe is expanding. We would not accept such a sweeping notion without testing its predictions. Two of these are illustrated in Figure 88, a sketch of the most important things that ever happened. The horizontal axis is time, t, since the birth of the Universe. The vertical axis, T, is its temperature or, more precisely, the temperature of its CBR radiation. As the Universe expands, T decreases with time, the blue line.

It is instructive to think of Figure 88 as a video and fast-run it backward for starters, from "now" toward the beginning of time. In terms of

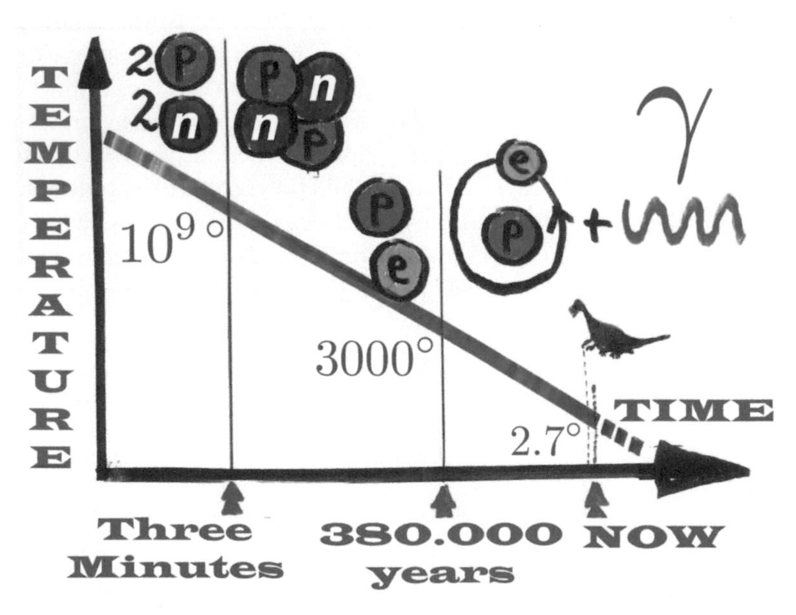

Figure 88 The most important things that ever happened. Chronologically from time zero: The synthesis of the primordial elements, the liberation of the Cosmic Background Radiation, and the extinction of dinosaurs.

Enjoy Our Universe: You Have No Other Choice. Alvaro De Rújula.

DOI: 10.1093/oso/9780198817802.001.0001

numbers of different particles, the Universe is essentially made of light plus a tiny fraction of ordinary matter, dominantly hydrogen (there are $\sim 6 \times 10^{-10}$ protons or electrons per photon[78]). In figuring out the past, we shall use the fact that we understand the behavior of radiation and matter. First example: Running time backward, we compress the stuff of the Universe, which consequently heats up like the compressed air in an old-fashioned pump used to fill bicycle tires, should you be old enough to have had this experience.

Momentarily forgetting the dinosaurs' demise[79] and similar ongoing disasters, the first thing that we see happening as we travel backward in time is that the temperature of the atoms (their kinetic energy) increases to the point where a hydrogen atom does not survive its collisions with others. Hydrogen breaks into its constituent protons and electrons. Practically simultaneously, this atomic plasma becomes non-transparent to the Cosmic Microwave Background radiation—by then compressed and heated to a temperature close to that of the Sun's surface. Matter and radiation establish thermal equilibrium.

As we visit the Universe back at a time when it was younger than a few minutes, the nuclei of atoms heavier[80] than hydrogen suffer collisions energetic enough to break them into their constituent protons and neutrons. We do not yet run the video to an even earlier time, when the non-uniformity of the Universe was allegedly generated.

The history displayed in Figure 88 becomes more interesting next, as we run the "video" forward in time. At an age of the Universe of a few minutes and a temperature of about a billion degrees, the pre-existing protons and neutrons coalesce into the nuclei of the *Primordial Elements*, the ones that were made then. Heavier nuclei are generated much later in supernova explosions and still heavier ones in neutron star mergers. The primordial elements are ^{4}He, "helium four" whose nucleus is $(2p, 2n)$, deuterium (p, n), ^{3}He $(2p, n)$, and ^{7}Li $(3p, 4n)$. Others, such as tritium $(p, 2n)$ are also made but are not stable.

The average abundances of primordial elements in the Universe have changed since they were made. Stars produce ^{4}He and destroy ^{7}Li.

[78] For this discussion we may forget about dark matter and the cosmological constant.

[79] To be precise, not all dinosaurs died. Some evolved into birds.

[80] One should—but generally does not—say more massive, instead of heavier. At the time the stuff of the Universe was homogeneous, the notion of weight made no sense.

Consequently, the extraction of the primordial element fractions (relative to hydrogen) is not so simple. Yet, the data agree with the predictions of the Big Bang Cosmology, as shown in Figure 89. Perfect agreement (but for ^{7}Li), considering in particular that the ratios extend over nine orders of magnitude. The theory of an expanding Universe is vindicated.

As we continue to play the history's video forward we get, when the Universe was about 380,000 years old, to the epoch that cosmologists call

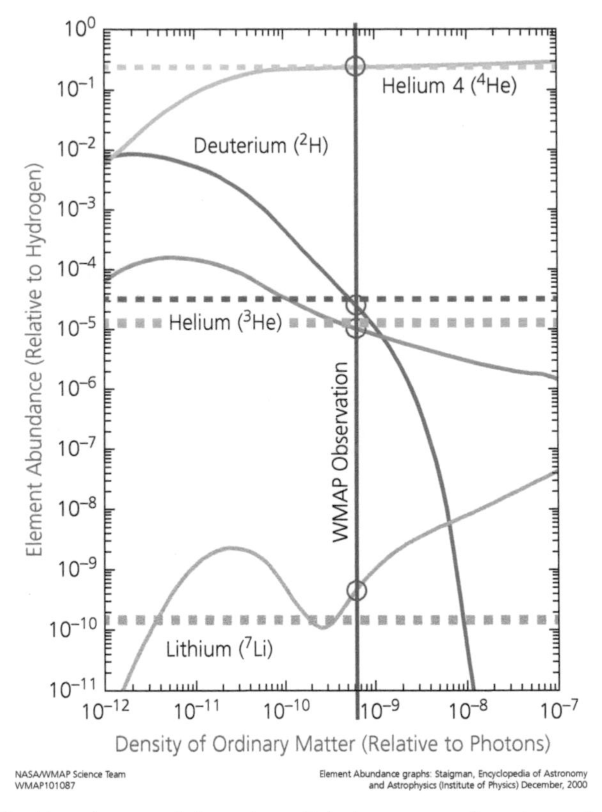

Figure 89 Abundances of the primordial elements as functions of the ratio of the number of protons and neutrons to photons (dominated by those of the Microwave Background Radiation). The red line is the measurement of this ratio by NASA's satellite WMAP. The continuous lines are the theory, as a function of this ratio. The circles are the consequent predictions. The dotted lines are the observed relative abundances. ©NASA.

"recombination." This refers to the time when electrons and atomic nuclei combined into atoms. They had never been combined before, but cosmologists favor confusing language,[81] competing in this with particle physicists.

In the mid 1940s, George Gamow, Robert Dicke, and others predicted that recombination would be accompanied by a "flash" of photons; the Cosmic Background Radiation. These ought to have survived to be observable today as the Microwave Background Radiation. The original estimates for its current temperature ranged from 5 to 50 Kelvin. This was a second impressive success of Big Bang Cosmology, the prediction of most of what there is. Indeed, in terms of numbers of observed particles, photons overwhelm any other (established) constituent of the Universe. We have not yet been able to directly detect cosmic neutrinos, another cosmic fossil of an expanding Universe.

The last event shown in Figure 88 is the disappearance of dinosaurs in our planet, 65 million years ago. This is "a minute ago" in cosmic times; the Universe had then 99.5% of its current age, $t_0 = 13.8 \times 10^9$ years. The times of recombination or nucleosynthesis are more than fairy-tale like: "Once upon a time...". They correspond to the Universe as a baby or a newborn: fractions 2.7×10^{-5} and 4.1×10^{-16} of t_0, respectively.

To all but very young Spaniards, the last dinosaur died on November 20th, 1975. Or so we used to think; In the twenty-first century it is no longer evident that the last of the dinosaurs was "Generalissimo" Francisco Franco.

[81] It is also said that they are often wrong but never in doubt.

26

Where is the Cosmic Antimatter?

Somewhat surreptitiously—without further precision or comment—we have said that the measured ratio of the number of protons plus neutrons[82] to the number of photons in the Universe is $\eta \simeq 6 \times 10^{-10}$, the red line in Figure 89. The ordinary matter of the Universe is made of atoms whose nuclei contain baryons (the proton and the neutron). The number η is called the *baryon to photon ratio*.

A brave and alert reader may have pondered the questions: Why have we been talking only about matter and radiation in the Universe? If matter and antimatter are nearly identical in their properties, why doesn't the Universe contain equal amounts of them? If it did, how come they did not annihilate each other resulting in a Universe with only light?[83] Summarizing: *How come I am here?*

A simple answer to the stated questions is: Somehow the Universe happened to be born in this peculiar way, with one baryon for every $1/\eta \simeq 1.7$ billion photons... and no antibaryons at all.

That was quite precisely the kind of answer that scientists consider ugly, unpalatable, and fully unsatisfactory. At a more scholarly level, these antimatter issues get even worse. Temperature is kinetic energy. At a sufficiently high temperature, larger for instance than the mass of the electron, the collisions between the constituents of the primordial plasma must have generated an equilibrium "constituency" of positrons. At a sufficiently high temperature, equally many electrons and positrons.

Up in temperature the argument above extends to antiprotons, antineutrons, quarks, and antiquarks of all types, and so on. And up to calculable factors not very different from unity the amounts of

[82] The number of protons and electrons is the same. Otherwise their charges would not add up to zero and there would be enormous (unobserved) electromagnetic forces competing with gravity in the Universe.

[83] The Universe must also contain a large but hard to detect population of neutrinos. We do not know whether or not the Universe is neutrino-antineutrino symmetric.

Enjoy Our Universe: You Have No Other Choice. Alvaro De Rújula.

DOI: 10.1093/oso/9780198817802.001.0001

photons and of matter and antimatter particles of each type must have been essentially the same. Not precisely identical numbers, though. Roughly speaking, for every 1.7 billion baryons there must have been 1.7 billion *minus one* antibaryons. If so, after *almost* all matter annihilated with antimatter, the teeny ratio η of matter to light is what happened to survive.

Surprise: The seemingly ugly last paragraph is the currently accepted theory. With one proviso to save it from the scientists' wrath: That the tiny asymmetry between the amounts of matter and antimatter was not "put there by hand," but was generated in an evolutionary way as the Universe evolved from a pleasing initial state with a vanishing baryon (and lepton) number. These numbers are defined as the quantity of particles minus antiparticles of a given type.

There are many concrete theories about how the matter excess of the Universe could be a result of evolution. They all share some key ingredients; for instance the obvious one being that the conservation of baryon number should not be an exact law of Nature (unlike the conservation of charge). None of these theories is much "prettier" than the others, or has made successfully tested predictions. Thus, they will remain unmentioned.

A not very appealing alternative to the cosmic domination of matter would be a Universe made of very large domains, some of them made of matter and others of antimatter. One thing we know for sure is that the visible Universe is not made this way. If that was the case, the inevitable annihilations at the domain boundaries would leave "ribbons in the sky"; unobserved scars in the Cosmic Background Radiation. This closes the antimatter case and brings us back to the CBR.

27

More on the Cosmic Background Radiation (CBR or CMB)

Thanks to a healthy international scientific competition and a succession of successful ground-based and satellite observations, measurements of the CMB have become astonishingly precise. A first example: The spectrum of the radiation (the intensity as a function of frequency) has a thermal shape so perfect that it would be inimitable in a laboratory experiment. The current CMB temperature is measured to be $T_0 = 2.7255 \pm 0.0006$ K (degrees Celsius above absolute zero). The spectrum obtained by the COBE satellite is shown in Figure 90.

Two entire-sky pictures of the CMB background radiation, gathered by the WMAP and Planck satellites, are shown in Figure 91. Their centers point to the galactic center. The plots cover the entire "celestial sphere," 90° up and down, 180° left and right. In the top panel the narrow line is the microwave radiation from the galaxy in which we are immersed.

The "dipole," The hotter blue and cooler red areas in Figure 91a, is due to the motion of our galaxy (and consequently our own motion) in the direction of the arrow, at a velocity $v \sim 10^{-3}\, c$ relative to the "absolute universal" reference system in which the CBR is almost isotropic (the same in all directions). Similarly, should you—like the gal in the figure—run on a windless day, you would feel an apparent "wind" on your face. The air molecules hit you harder there.

Shown in the lower panel of Figure 91 is the microwave sky after the galaxy and the dipole are subtracted. The remaining anisotropies (the bluer or redder spots) are now only at the tiny level of $\sim 10^{-5}$ of the mean $T_0 \sim 2.7$ K; the CBR is almost identical from all directions. The observed microwave photons have been redshifted by a factor $z_R \sim 1100$ from the time of recombination when they were emitted until now.

There are two ways to analyze how a ring vibrates. The easy one is to kick it, record its sound, and study its different frequencies. One would simply hear a "fundamental" frequency and its multiples or

Enjoy Our Universe: You Have No Other Choice. Alvaro De Rújula.

DOI: 10.1093/oso/9780198817802.001.0001

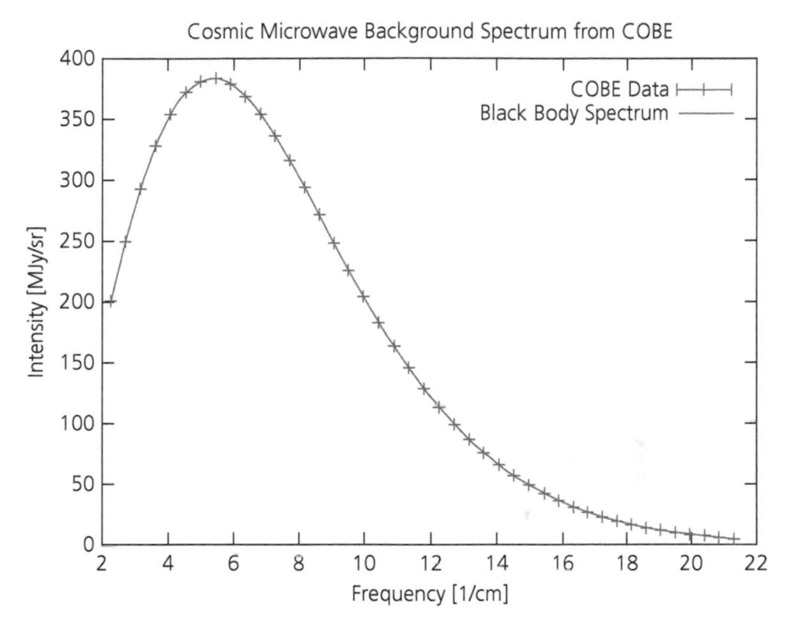

Figure 90 The perfectly thermal shape of the CBR spectrum. "Thermal" refers to the emission by an ideal "black body" at a given temperature. ©NASA.

"harmonics." The main frequency corresponds to the entire ring getting stretched or squeezed in some direction, as in Figure 92a. The harmonics correspond to vibrations with more crests and troughs, as in Figure 92b.

Another way to analyze the ring's vibrations is by eye, rather than by ear: Take a snapshot and study its structure. Looking at segments of the ring of various lengths (or opening angles from the center), how much do they deviate from a ring at rest? An answer to this question is the *power* as a function of *angular scale*. The power at a particular angular scale is computed by averaging the squares of the deviations of the ring's shape—the blue or red lines in Figure 92(a) and (b)—from the undeformed ring—the dashed lines.

The analysis of the anisotropies of the snapshot[84] of the CBR as a function of their angular size is two-dimensional, but otherwise similar

[84] The maps of the CBR are indeed "snapshots." The CBR does not come from an ideally thin surface, but from a progressively less transparent "orange skin," some 10,000 light-years thick. Roughly speaking, in 10,000 years the CBR light will come to us from 10,000 light-years further away and a snapshot of the CBR will be different.

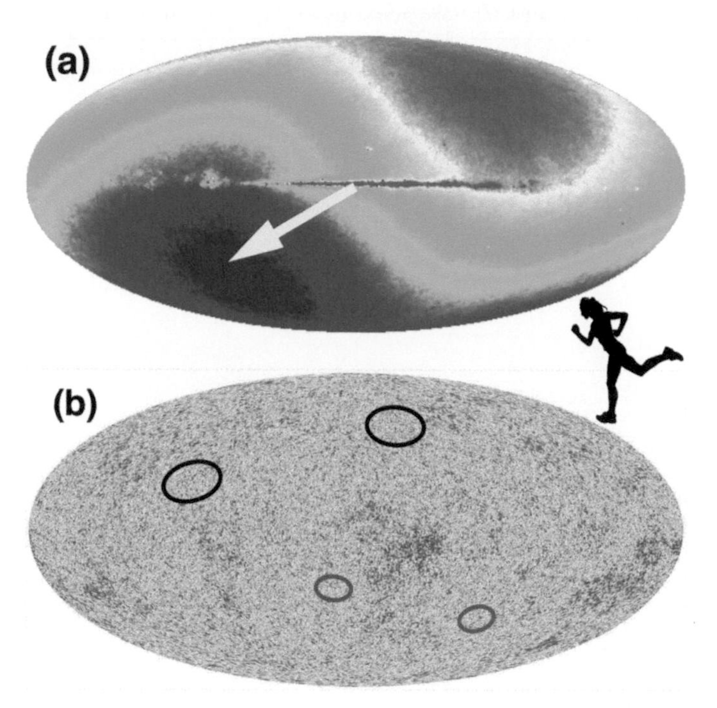

Figure 91 Maps of the CBR. In (a) the dipole is dominant and the galaxy is visible ©NASA/WMAP. In (b) they are subtracted, revealing the true intrinsic CBR anisotropies, ©ESA and Planck Collaboration, analyzed in regions of various angular scales, of which two examples in two directions are shown. Runner: See text, from https://openclipart.org/detail/259731/female-runner-2.

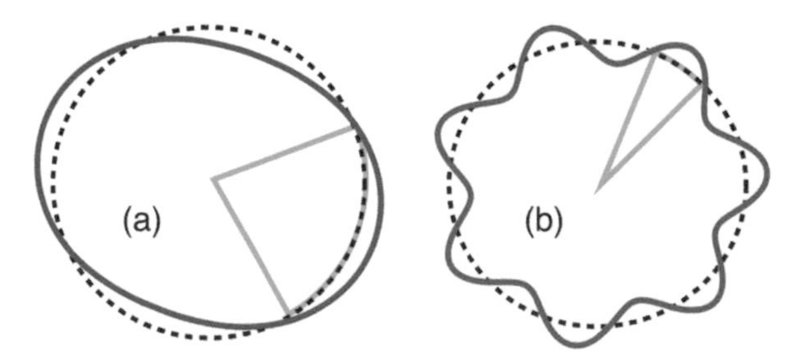

Figure 92 Dashed lines: A non-vibrating ring. Green arcs: Angular apertures. (a) The fundamental vibration. (b) The sixth harmonic.

to the ring's example in the previous paragraph—it has to be slightly more sophisticated to take into account that the CBR irregularities are randomly distributed and not perfectly organized, as in Figure 92(a) and (b). The angular scales (or sizes) trace circles around a given direction in the sky, and are moved around to cover all of it and to average the power as a function of the diverse angular sizes. Two such sizes are shown in Figure 91, a black and a red one, which in the spherical projection of Figure 13 would look like circles, not ellipses. The measured power of the CBR as a function of angular size is shown in Figure 93.

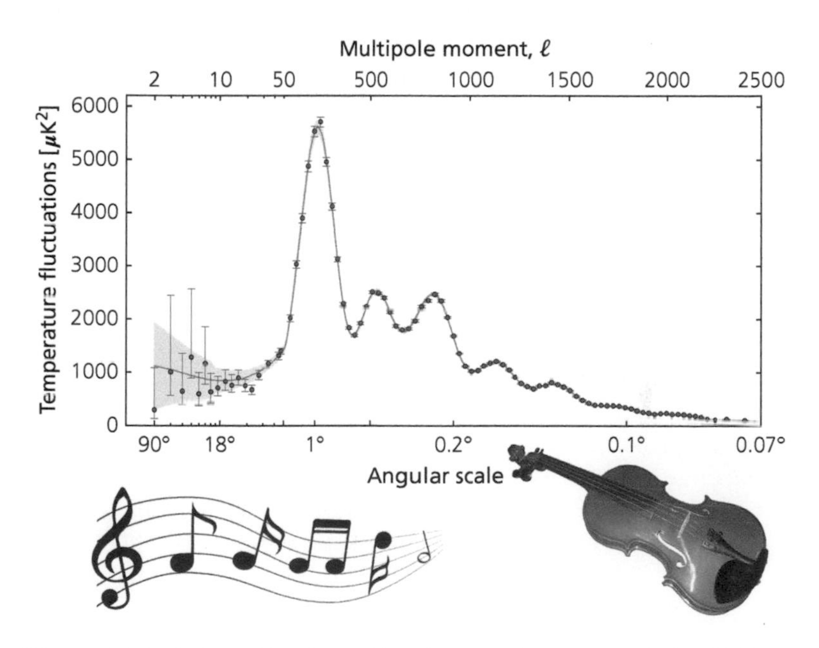

Figure 93 Red dots: Planck-satellite data on the power of the CMB as a function of the angular scale. Green line: A fit to the ΛCDM model, described in some detail in the first part of Chapter 29. ©ESA and Planck Collaboration. Musical notes: http://www.publicdomainpictures.net/view-image.php?image=127060&picture=musical-notes, Violin by Brunothg, https://upload.wikimedia.org/wikipedia/commons/thumb/3/33/Violin Geige.png/2048px-Violin Geige.png.

Analysis of the CBR anisotropies

If you listen to the sound of a musical instrument, you do not need to look at it to guess what instrument it was. A violin as in Figure 93, for instance. Had it been a flute, your ear could even discern whether it was made of metal or wood. The sound analysis made by your brain is mainly based on the relative intensities of a fundamental note and its harmonics. The analysis of the anisotropies of the CBR, which are also due to vibrations of matter, is quite analogous.

It is difficult to calmly express how extraordinary the consequences of an understanding of Figure 93 are: They are one of the most wonderful summaries of our knowledge ... and of our ignorance. The key point is that from the sizes and positions of the fundamental vibration (at an angular scale of $\sim 1^\circ$) and its harmonics (at $\sim 1/2^\circ$, $\sim 1/3^\circ$, etc.) it is possible to deduce what the Universe is made of and how "flat" space is. And that is where the surprises abound.

The detailed analysis of Figures 87 and 93 requires a model of the Universe. The "Old" Big Bang Cosmology we have outlined has flaws that are solved in *inflationary* models, as we shall discuss. To whet the appetite for them, let us first look at what they imply concerning the global properties of the Universe, summarized in Figure 94 and Table 1.

The cosmic pie on the left of Figure 94 is meant to dramatize that, in terms of average density per unit volume, the Universe is mainly made

Table 1 Approximate values of the parameters describing the Universe in the ΛCDM model. Actual densities, ρ, are expressed as ratios to the critical one, for example for the vacuum energy $\Omega_\Lambda = \rho^0_\Lambda/\rho^0_c$. Ω_m, Ω_{om}, and Ω_r refer to total matter, ordinary matter, and radiation, respectively.

H_0	**67 km/s/Mpc**	$\lvert\Omega_0 - 1\rvert$	$< \mathbf{0.007}$
ρ^0_c	$\Leftrightarrow 5\,\text{Hydrogen}/m^3$	Ω^0_Λ	0.7
t_0	13.8×10^9 years	Ω^0_m	0.3
T_0	2.7255 °K	Ω^0_{om}	$\sim \Omega^0_m/6$
z_R	1100	Ω^0_r	$5.5\,10^{-5}$

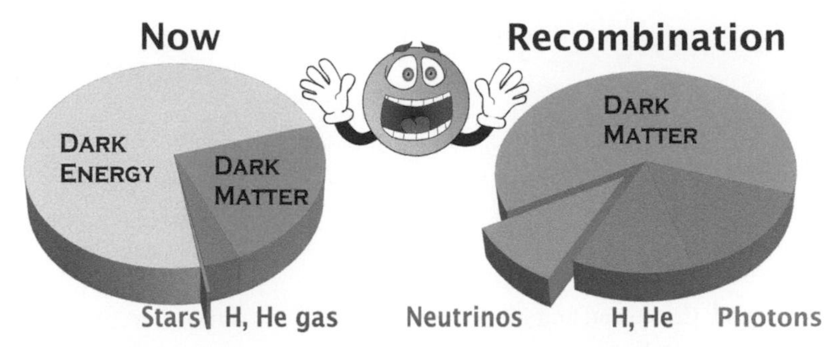

Figure 94 The energy-density pie of cosmic essences, now and at recombination time. The densities evolve differently as the Universe expands. The "recipe" to make our Universe is indeed astonishing.

of substances we know very little about: Dark energy and dark matter. And the stuff we are made of (ordinary matter) is mainly in the form of hydrogen and helium gas, stars are only a small fraction of it, planets are negligible.

In Table 1, and in the usual notation, the "0" indices refer to "now." Only the redshift at recombination, z_R, alludes to the past. We have already discussed the Hubble constant (constant in direction, not in time), the age of the Universe, t_0, and the CBR temperature, T_0.

Our understanding of the Cosmos is based on Einstein's equations[85] in Figure 1. They imply that the contents of the Universe shape the structure of its space-time. There is a particular energy density, called *critical*, whose current value is roughly equivalent to that of a Universe with five hydrogen atoms per cubic meter, but nothing else. The ratio Ω of the total average density in all kinds of "stuff" to the critical density is measured to be compatible with $\Omega = 1$ within very small errors.[86] For such a critical value, space (at a fixed time) is flat.[87] For $\Omega > 1$ ($\Omega < 1$) the geometry of space would be closed (open), as in Figure 95.

[85] More precisely, on their isotropic, homogeneous solutions, called FLRW, after Friedman, Lemaître, Robertson, and Walker, who first found them.

[86] Different observables, such as the CMB and the Hubble plots, agree on their consequent results for quantities such as H_0 and Ω_Λ. With characteristic dramaturgy cosmologists call this minimal scientific requirement "the cosmic concordance."

[87] In a flat space the sum of the angles of a triangle is 180°. For closed or open geometries it is more or less than that, respectively. Thus, one need not go "out" of the space—into the white of Figure 95—to figure out its geometry.

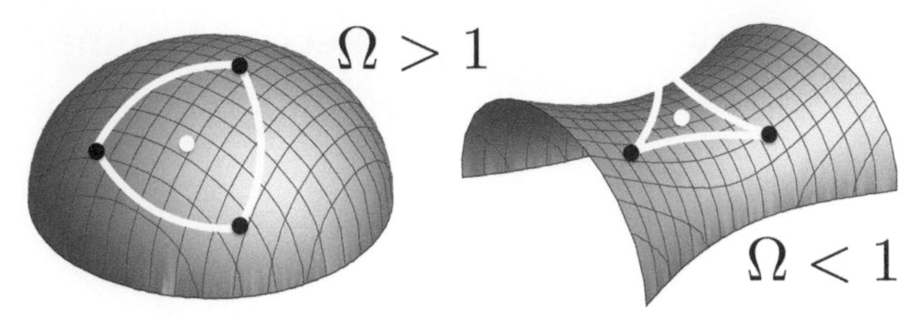

Figure 95 Closed and finite and open and infinite two-dimensional surfaces.

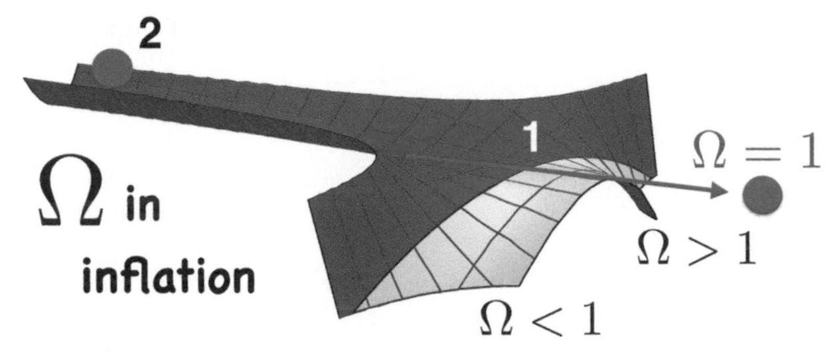

Figure 96 Fate of the normalized total density Ω in a universe with $\Omega = 1$ along the "critical" blue line. Launched from position "1" the ball falls to an $\Omega \neq 1$ side. Unless the bowler throws the ball, from the position "2," along a long valley that places it precisely at $\Omega = 1$ during an "inflationary" period (in red).

The fate of a universe is akin to that of the blue bowling ball in Figure 96. Suppose we throw it to the right from the position "1" with the intention of not having it fall to one or the other side of the blue *critical* line at the top of the yellow vault, a bowler's lane designed by an evil character. That would require an extraordinary bowler's skill, for a minimal inaccuracy would result in the ball falling to one or other side.

Thank goodness, somehow, our Universe is critical. The reason is that if it weren't we would not have a chance to exist. A minutely overcritical universe (at the time of the production of the primordial elements, for instance) would currently have, like the ball in Figure 96, ended up well into the $\Omega > 1$ side. This implies that it would have suffered, long ago, a catastrophic Big Crunch. A minutely under-critical universe would have expanded at a fast rate, ordinary matter would

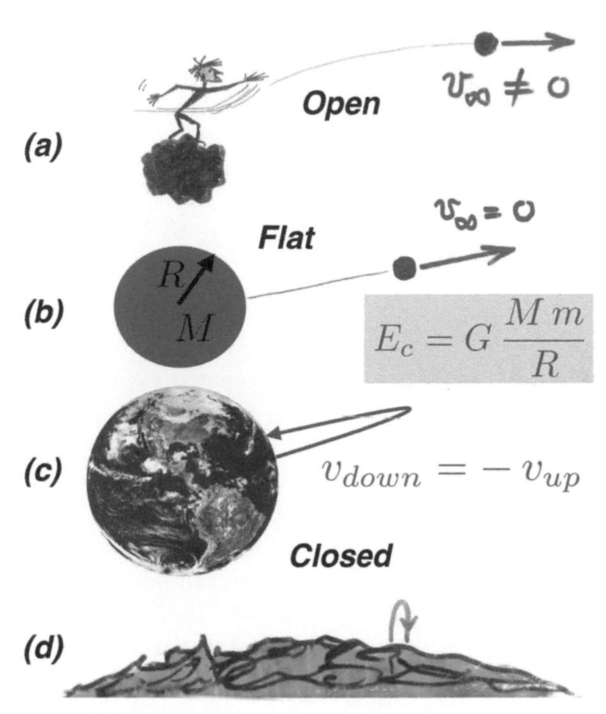

Figure 97 (a) A subcritical asteroid or open universe, $\Omega < 1$. (b) A critical mini-planet or flat universe, for which the ball's kinetic energy equals its gravitational energy at the surface, and $\Omega = 1$. (c) The over-critical planet Earth, or a closed universe, $\Omega > 1$. (d) A larger rocky planet. Earth by Reto Stöckli, Nazmi El Saleous, and Marit Jentoft-Nilsen, ©NASA GSFC.

not have managed to condense in galaxies, stars, and planets... and we would not be here, either.

Another way to emphasize how ludicrous the idea of a critical universe appears to be is illustrated in Figure 97. A baseball player throws a ball of mass m WEAHM (With Exactly All His Might). If he is standing on a small asteroid the ball will escape its gravity and after an infinite time it will still be flying away with a velocity v_∞. If he is standing on Earth the ball will fall back, even if thrown WEAHM. The ball may stop at $v_\infty = 0$ only if thrown from a planet with a mass M and radius R such that the gravitational energy of the ball on its surface ($G\,M\,m/R$) is *critical*: Exactly equal to the ball's kinetic energy when the player throws it WEAHM, a preposterous coincidence.

28

Problems with the "Old" Big Bang Theory ★★

Why should the Universe be born or be prepared, as in Figure 96, in such a way as to be critical? So that we can pose the question, say some.[88] Criticality is a first conundrum in the "old" Big Bang Theory. A second one is "causality": How come the Universe, pictured at the time of recombination, is as uniform as we see it in Figure 91?

To summarize the following puzzling discussion on causality: The observed MWBR of Figure 91b has the same temperature (to about one part in 10^5) in all directions. But the regions that could have reached a thermal equilibrium by the time of recombination (when the radiation was emitted) subtend only a fraction (about 1/70) of a complete circle in an angular scale. They "should" have unrelated temperatures. But they do not.

To discuss causality it is useful to consider a closed universe with two space dimensions, pictured in Figure 98. This universe is the 2D *surface* of the depicted spheres, nothing above or below it actually exists; the third dimension, illustrated by the arrows in Figure 98a, is only a (very useful) "mathematical" manner of speaking. It allows us, for instance, to think of the expansion of the Universe as a function of time as an increasing $a(t)$, the "radial scale." The "angular distance" χ_{12} between two galaxies, pictured as stars, is kept fixed as the "balloon" swells; it is the space between the galaxies that gets stretched.

In Figure 98b we see this universe "now," at time t_0. An observer O placed at the green dot position sees the CBR emitted in the past from what is now the purple circle. Information has traveled to O from any place in this circle at the speed of light, c, subtending an angular distance χ_0, the maximum one between two places that can be "casually

[88] Somewhat jokingly, I used to hope, some cosmologists discuss an *Anthropic Principle*, the strong version of which is that things are the way they are so that we can also be here to see that things are the way they are.

Enjoy Our Universe: You Have No Other Choice. Alvaro De Rújula.

DOI: 10.1093/oso/9780198817802.001.0001

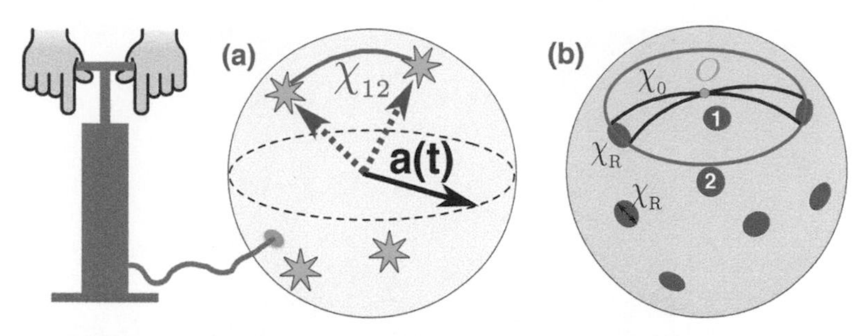

Figure 98 A two-dimensional universe expanding with a scale $a(t)$, as if something was blowing it up. (*a*) The *angular distance* χ_{12} between two distant galaxies does not change with time, but their distance does, proportionally to $a(t)$. In (*b*), the inside of the red spots (in an old Big Bang cosmology) was causally connected at recombination, when the CBR originated. Each of them could have its own temperature. The observer O now sees the CBR coming from the much larger, noncausally connected mauve circle. Why is it all at the same temperature?

connected" at time t_0; that is, two locations that can have exchanged information at a speed smaller or equal to c. The CBR light from the red spot "1" arrived to O in the past, light from "2" will arrive in the future. Once again, we stumble upon the notion that a universe is bigger than what we see.

From the birth of our toy 2D universe to the time, t_R, when its matter recombines into (2D, flat) atoms and its CBR is emitted, light could only have traveled for a certain angular distance, χ_R. Since t_R is much smaller than the current age of the universe, t_0, the angular distance χ_R is much smaller than χ_0. The matter inside a red spot in Figure 98b was causally connected at the time of recombination. Spots of that size have the "right" to emit CBR that O sees at a uniform temperature. But two distinct spots, like the ones O sees in two opposite directions, *have no way to have decided to be at essentially the same temperature*. And they are observed to be. That is, now in detail, the causality problem.

Much as the criticality or flatness of Figures 96 and 97, the question of causality brings about the very unsavory issue of initial conditions. To be born causally connected, the universe of Figure 98 must be uniform from the start: At t very close to zero, its entire space would have to have the same properties; even if, then, any region "having the right" to be causally connected would be a tiny fraction of the entire primordial

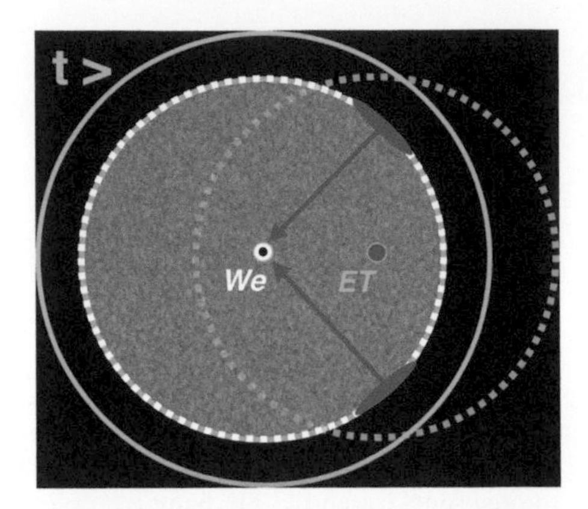

Figure 99 Our very large Universe. In black, a small portion of what we have not yet seen. The CMB currently reaches us from the white-dotted line. Inside it, in gray, the tiny visible fraction of the Universe. ET sees a different region, within the blue dotted line. Future cosmologists on Earth will see a bit further, to $t_>$. In the old Big Bang model, different red patches were *not* causally connected when the CMB was emitted.

Universe. Arbitrarily fixing such an extremely peculiar initial state is, for a physicist, disagreeable, irksome, annoying, irritating, displeasing, distressing, awful, dreadful, objectionable, offensive, obnoxious, repulsive, revolting, disgusting, distasteful, nauseating, and other synonyms of "unsavory" cited by the OED.

The unfamiliar concepts we face when understanding *a* universe get only more weird as we study *our* Universe. Since its measured Ω is so close to one, the Universe's space is nearly flat. For Ω bigger than—but very close to—unity, the Universe is nearly infinite. For $\Omega < 1$ it is infinite, a notion of which only philosophers (not scientists) are afraid.

A 2D slice of the space of the Universe is illustrated in Figure 99. The part "We" see is gray, what we have not seen is black and extensive. Sometime in the future, we shall see up to the green "horizon" $t_>$. An "ET" currently living at the blue spot sees a CBR originating in the blue-dashed region. The red spots in our CMB horizon were causally connected at recombination. Again, only regions of their size or less may have been causally connected so as to emit light of the same temperature, contrary to observation. Back to the drawing board, as the scientific method of Chapter 4 requires.

29

Inflation

To answer the questions we have been posing, all we need is to find a good reason why:

- The Universe is flat—or critical, with $\Omega = 1$, very precisely.
- The CBR is so uniform—so similar in all directions.

Killing these two birds with one stone is (a posteriori!) extraordinarily simple. And, as a bonus, the solution called inflation entails beautiful predictions that turn out to be right. There is no particular model of inflation that is more predictive, successful, or elegant than others. All of them require some choices to be made so that their results are acceptable. Therefore, we shall not discuss a particular model, but the features they all share.

If the average energy density of the Universe is dominated by the vacuum contribution, Λ, the solution of Einstein's equations of Figure 1 is *inflationary*. What this means is that the scale $a(t)$ increases at a fast and accelerating rate, exponentially with time.[89] When the idea was developed[90] in the early 1980s it sounded suspiciously farfetched. But nowadays we know that the Universe is inflating right now! This is implied by the cosmic pie on the left of Figure 94 and by the direct rate-of-expansion measurements of Figure 87: The vacuum energy density is currently dominant.

The relative contributions of the cosmic substances to the density of the Universe vary with time,[91] see the differing pies of Figure 94. The Inflationary Paradigm is the contention that, for a tiny but finite

[89] For matter (radiation) dominance, $a(t)$ increases much more slowly, like $t^{2/3}$ ($t^{1/2}$).

[90] By Alexei Starobinsky, Allan Guth, Andrei Linde, Paul Steinhardt, Viatcheslav Mukhanov, and others. The paternity is shared and hard to weigh.

[91] The matter-density contribution ρ_m is mass/volume. Mass does not change and volume increases like $a(t)^3$, so that $\rho_m \propto a^{-3}$. The energy of radiation itself is redshifted as $1/a(t)$, so that $\rho_r \propto a^{-4}$. The cosmological constant contribution is… constant.

Enjoy Our Universe: You Have No Other Choice. Alvaro De Rújula.

DOI: 10.1093/oso/ 9780198817802.001.0001

amount of time (some 10^{-32}s), the primordial Universe was "vacuum-dominated," consequently undergoing a temporary inflationary phase. The (twice) inflationary expansion history of the Universe is illustrated in Figure 100.

In the primordial inflation the Universe expanded by very many orders of magnitude. The tiny fraction of it that is visible to us, like a piece of confetti glued to a giant sphere, is practically flat. According to General Relativity, flat and critical ($\Omega = 1$) are equivalent. So, inflation works in a way analogous to the long funnel preceding the bowling alley of Figure 96.

Carefully comparing the "old Big Bang" results of Figures 98b and 99 with the inflationary ones of Figure 100 one notices a crucial difference (did you?). In the later figure, the causally connected red patch is bigger than the visible universe and not the other way around. This is a consequence of General Relativity, not a drawing error. Indeed, inflation solves the causality problem, thus explaining why the CMB is so uniform.

I did not say *how* inflation solves the mentioned problem. The excuse is that the notion of causality is a good deal less obvious in an expanding universe described by General Relativity than it would be in a static one described by Special Relativity. A precise explanation would be

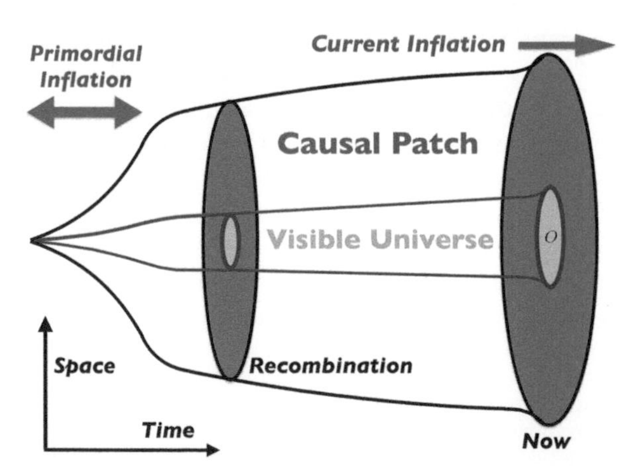

Figure 100 The two inflations of the space of the Universe. In between them, a more conventional expansion. A causally connected (red) domain is much bigger than the (gray) spot representing the visible Universe.

fairly hard to follow[92] and what is worse, boring. Should you be a bit mesmerized by all these considerations on spaces evolving in time, see Figure 101.

The CBR is not perfectly featureless and the matter of the Universe is not distributed as a uniform gas would be (only very large portions of the Universe look very much alike, having—for instance—the same average density of matter). At first sight, inflation would get rid of the CBR anisotropies, as well as of galaxies, stars, planets, and people. In fact, as we shall see anon, it triggers their existence.

Figure 101 *The sleep of reason creates monsters,* an etching by Francisco de Goya, adapted in connection with a statement by Einstein: *People slowly accustomed themselves to the idea that the physical states of space itself were the final physical reality...* For an earlier commentary wherein you may find your grandfather, see https://history.aip.org/exhibits/einstein/ae27.htm

92 A very rough explanation is the following. At sufficiently large angular distances—and not violating causality—*space itself* can grow faster than light. During inflation initially causally connected domains do so, internally "freezing" their causal information. After inflation these domains grow less fast than the "horizon" from which light can reach an observer. Our horizon is well within a causal domain.

Inflationary predictions

In inflationary models of the Universe, the key actor is a scalar (spin 0) time-dependent field, $\Phi(t)$, called *the inflaton*, somewhat analogous to the Higgs field discussed in the beginning of Chapter 22. During inflation, $V(\Phi)$ (the energy density of Φ) acts as a temporary cosmological constant and inflates the Universe. As time elapses, the field $\Phi(t)$ "rolls down" like a ball, as in Figure 102, toward its minimum at $\Phi = 0$ and $V(0) = 0$ (by now you may have noticed that physicists are fond of rolling balls).

Before it settles down, $\Phi(t)$ oscillates, spending its energy in creating all the other particles of dark and ordinary matter that permeate the Universe. The behemoth entropy of the Universe (the logarithm of the number of states of all of its particles) is developed in an evolutionary manner from a single primordial state of the inflaton, an initial state of pleasantly maximal simplicity. The process of particle generation is called *reheating*, even if nothing had to be hot before (once again, the confusing lingo of cosmologists).

The nicest prediction of inflation concerns the origin and nature of the observed CMB anisotropies. They are generated by quantum fluctuations of $\Phi(t)$ and stretched to cosmic scales during inflation. They

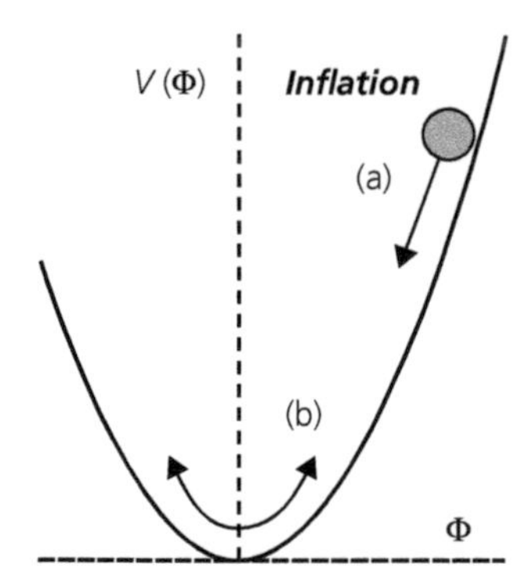

Figure 102 An inflaton field Φ first rolls down (a) and then oscillates (b) during inflation and reheating. $V(\Phi)$ is an example of "inflationary potential" describing how Φ interacts with itself.

are maximally random (or *Gaussian*), they do not have a specific scale and they have a predicted polarization,[93] all as observed.

The peaks in Figure 93 are *acoustic* in the sense that they relate to the speed of sound vibrations in the primordial plasma of ordinary matter.[94] The 1° angular scale of the main harmonic in Figure 93 corresponds to the distance sound has traveled up to the time of recombination, the size of the largest blobs of plasma that may have vibrated coherently before the plasma became a more inert gas. These once-vibrating spots are the largest objects in the Universe with an identity: The largest *things*. And they are generated by quantum fluctuations of the inflaton's field. *Quantum mechanics describes not only the behavior of the smallest things (elementary particles), but also of the largest!*

High time to clarify what the mysterious acronym ΛCDM in Figure 93 and Table 1 means. It refers to a universe dominated by a cosmological constant (Λ) and **Cold Dark Matter** (CDM). Dark matter interacts with ordinary matter only via gravitation. "Cold" means that the hypothetical particles that constitute the DM have a temperature (kinetic energy) negligible relative to their mass (times c^2, if you forgot that $c = 1$ in natural units).

Cosmologists often present results such as the green line of Figure 93 as a *prediction* of inflation. It is not. It is actually a fit to the ΛCDM model with half a dozen parameters fixed by the fitting procedure. Moreover, the general trend of the data was predicted by Ted Harrison, Jim Peebles, and J.T. Yu in 1970, well before the invention of inflation. None of which subtracts from the considerable success of the Inflationary Paradigm.

[93] This is an *E-mode* polarization. Also predicted but not yet convincingly observed is a *B-mode* polarization, related to gravitational waves generated during inflation.

[94] In a 3D relativistic plasma the speed of sound is $c/\sqrt{3}$. These two 3s are the same.

30

Limitations and Peculiarities of the Standard Models

The ΛCDM model of cosmology, an "exercise" in General Relativity, has become "Standard"; in a sense somewhat similar to the Standard Model of elementary particles and their interactions. Critics of the ΛCDM model point out that the initial conditions, for instance on the actual shape of the inflaton's potential of Figure 102, have to be "tuned just so that the model works" and accommodates, among other things, the small observed sizes of the temperature fluctuations in Figure 93. The fact that we happen to live in an epoch in which the fit to these fluctuations results in a *recipe to cook a universe* such that the amount of the various ingredients (vacuum energy, dark, and ordinary matter) are *currently* similar, as in Figure 94, may also be regarded as "unnatural." Whatever that means.

The particle's Standard Model also has features that are considered unnatural. The main one is that it does not have an automatic mechanism to make the Higgs boson as "light" as it is.[95] The range of masses of elementary fermions is also surprisingly "unnatural": The top quark mass is $\sim 3.4\,10^5$ times the electron's. And the model has even more free parameters (to be fit to observations) than the cosmological model.

Subsequent to the discovery of the Higgs boson, it is often stated that the particle's Standard Model is complete. Not true. The model predicts the existence of a neutral, spin zero, light, very weakly interacting particle called the *axion*, a possible candidate constituent of dark matter. The axion has not yet been discovered, in spite of many varied and ingenious "hunts" for it. One of them is shown in Figure 103.

[95] This is in comparison to the Planck mass, $M_P = 1/\sqrt{G} \approx 1.22 \times 10^{19}$ GeV, some 10^{17} times the Higgs mass. The model would "naturally" drive up the Higgs mass to be similar to Planck's one.

Enjoy Our Universe: You Have No Other Choice. Alvaro De Rújula.

DOI: 10.1093/oso/9780198817802.001.0001

Figure 103 CERN's CAST "magneto-telescope," searching for axions emitted by the Sun. In the magnet, axions would transmute into detectable photons. Courtesy of CERN/CAST.

It has to be inescapably admitted, in spite of all the above, that the Standard Models—incomplete and "unnatural" as they may be regarded to be—offer a fantastically accurate description of Nature. Their limitations are only room for further progress in our understanding, methinks.

31

The Discrete Symmetries of the Fundamental Interactions ★★★

Rotations are "continuous symmetries." One can rotate a circle by an arbitrary number of degrees, it still looks the same, it is "invariant" under rotations. A square is only invariant if you turn it a certain number of degrees: 90,180, and so on. Such a discontinuous symmetry is called "discrete." We shall discuss three of them and some combinations thereof.

Somewhat surprisingly, the discrete symmetries of the Standard Model are not so easy to understand; they involve unfamiliar notions connected with relativity and quantum mechanics. The reader not immediately jumping to the next chapter—and daring to read this one—may find it more or less grueling, but will definitely deserve an A^{++}.

Parity is the operation relating an object to its mirror image, like the characters in Figure 104. If you are left-handed, could your right-handed reflection in the mirror also be a functioning you? The answer is affirmative. That is because the weak interactions play no role in the way you "function" at the moment.[96] The rest of the fundamental forces are *parity conserving*, while the weak interactions are maximally *parity violating*.

Some observed weak decays of charged kaons ($s\bar{u}$ and its antiparticle $u\bar{s}$) appeared to be impossible if parity-symmetry was assumed, which was done up to the mid 1950s, as if it went without saying. Tsung-Dao Lee and Chen-Ning Yang explained the kaon decay puzzle and predicted other very specific effects by assuming that weak interactions violated parity, a revolutionary step at the time. Chien-Shiung Wu and collaborators performed the first experiment proving Lee and Yang right.

[96] Our amino acids are left-handed and our sugars right-handed. Whether parity violation by the fundamental interactions played a triggering role in this is not yet settled.

Enjoy Our Universe: You Have No Other Choice. Alvaro De Rújula.

DOI: 10.1093/oso/ 9780198817802.001.0001

Figure 104 Madame Wu, T.D. Lee, and C.N. Yang and their mirror reflections. You cannot tell who the "real" ones are, but Nature can. Credits: Wu and Lee, wikipedia. Yang, nobelprize.org.

And here comes the most up-hill part of this chapter. Remember that the number of stars after a chapter's title indicates its relative difficulty. And this one has three stars! What it tries to explain is that the notion of particles and antiparticles may be more involved than what everybody recalls, namely that elementary particles and antiparticles have opposite charges (or opposite colors) and one particle may interact with its antiparticle to annihilate each other, converting their total energy (including their rest mass) into the energy of other lighter particles—like in the reaction in which an electron–positron pair transmutes into photons, as in Figure 23. The news to come is that neutrinos, which are neutral, may differ from antineutrinos in another way, not to do with charges, but with the way they move and spin.

"Dirac's" quantum field of an electron describes four objects: Spin up or down electrons and spin up or down positrons (a spin 1/2 particle may *only* have its spin point *up* or *down*, when measured along a given direction, marvels of quantum mechanics). For a spin-up particle traveling in the "up" direction with an energy much much bigger than its mass, the projection of its spin along the direction of motion is called *chirality*, from the Greek for "hand."

A left-handed particle is one that travels in the direction of the thumb of your left hand and spins in the direction pointed by the rest of the fingers of the same hand. Left- and right-handed electrons are thus depicted in Figure 105. A very good baseball pitcher would think most of this goes without saying. If he can shoot right- or left-handed balls, as we shall see, the batter is in real trouble. By the way, the photon in Figure 26 are also either right- or left-handed.

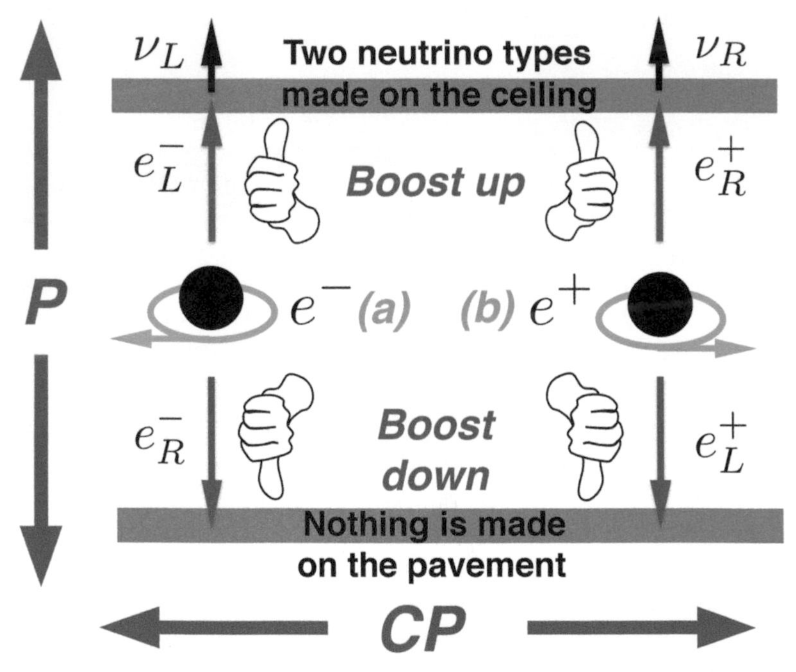

Figure 105 The behavior of spinning e^-s and e^+s as they are ultra-relativistically boosted up or down. Only left-handed e^-s and right-handed e^+s make neutrinos (top of the figure). You may have to turn the page around or to stand upside down to check the handedness of the particles hitting the pavement. The upper and lower halves of the figure are related by *P*, the left and right halves by *CP*.

Take an electron spinning as in the entry *(a)* of the busy Figure 105 and boost it to the "ceiling" so that it arrives there *almost* at the speed of light: It is *almost* in a pure left-handed state.[97] Or boost it to the "pavement" in a similar fashion. Do the same for a positron from its position at rest in *(b)* of Figure 105. Arriving to the ceiling or pavement, how do these particles weakly interact with matter and make neutrinos? The answer is parity violating. And astonishing.

In a mirror an arrow pointing to it appears to point in the opposite way, like the blue arrows in Figure 105, which illustrate the opposite velocities of the e^- and e^+ boosted up or down. But if you look at the turning hands of a clock—the one for seconds, if you are impatient—you see them in the mirror turning the same way. The up- or down-moving e^- or e^+ in Figure 105 spin the same way but move in opposite directions. The upper and lower parts of the figure are thus related by a parity transformation.

On the ceiling of Figure 105, left-handed electrons make left-handed neutrinos, ν_e^L, via the weak interaction process $e_L^- p \rightarrow \nu_e^L n$. Similarly, right-handed positrons make right-handed neutrinos, $e_R^+ n \rightarrow \nu_e^R p$. Only ν_e^L and ν_e^R have been observed. Compared to the number of electron states, two possible neutrino types are missing, the ones that one might have expected to be made at the pavement of Figure 105. We do not know whether the extra guys are "sterile" (they only interact with gravity), or they simply do not exist. Back to our friendly batter, he would miss half of the electron (or positron) balls, depending on how the pitcher made them spin. And he may not even know why! Do not fear for your favorite team, quantum pitchers such as the one in Figure 106 are not yet around.

In only ν_e^L and ν_e^R exist, they would be a particle–antiparticle pair of "Majorana" neutrinos.[98] This option opens a deep new kind or relationship between matter and antimatter, in the laboratory and perhaps in the cosmos as well.[99] The fact that a Majorana neutrino may, depending on the relative orientation of its spin and its motion, create a particle or

[97] Handedness is the high-energy limit of *helicity*, the projection of spin along the direction of motion. Helicity can be flipped by an observer speeding up to overtake the observed particle. Handedness is observer independent: relativistically invariant.

[98] After Ettore Majorana, the brilliant young Sicilian physicist never found after his mysterious disappearance, when crossing from Palermo to Naples . . . or later?

[99] Hypothetical Majorana particle–antiparticle pairs may be involved in generating the excess of matter over antimatter in the Universe, the subject discussed in Chapter 26.

Figure 106 Quantum baseball. The young Majorana always missed returning a neutrino when hitting a right-handed electron ball. Not so for left-handed ones. Charlie Brown only tries to bat sterile neutrinos, with known results. Here the balls and not the pitcher are right- or left-handed.

an antiparticle is quite astonishing. The nature of neutrinos is an open question, discussed here as an elaborate example of the many things we know that we do not know. Yet.

The attempts to determine whether neutrinos "are Majoranas" exploit a process called *double β-decay*. One of these experiments is shown in Figure 107. Some atomic isotopes spontaneously double β-decay via a process in which two neutrons of their nuclei, in a single step, decay into two electrons, two neutrinos, and two protons of a daughter nucleus; the one two steps up in the Periodic Table. This rather conventional process is illustrated in the entry (*a*) of Figure 107.

If it is not massless, a right-handed neutrino may be brought to rest, with its spin pointing in the direction in which it was previously moving. If boosted in the opposite direction, it travels as a left-handed neutrino. The process illustrated in the entry (*b*) of Figure 107 is akin to the ones in Figure 105, except that the ceiling and pavement are replaced by the upper and lower neutrons and the particle moving between them is a neutrino, not an electron or positron. At the "ceiling" an e^- is made alongside a right-handed neutrino, as is always the case. If the "descending" right handed neutrino is Majorana, it propagates in such a way that its mass can "flip it" to look like a left-handed neutrino. Hitting the lower neutron, the left-handed neutrino makes another e^-, as usual.

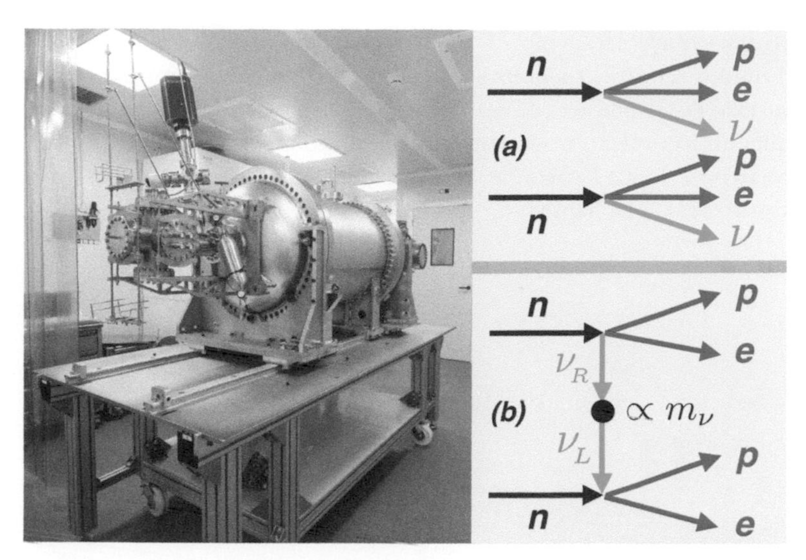

Figure 107 LSC: a tabletop experiment attempting to reveal a hypothetical Majorana nature of neutrinos and to constrain their masses. Located at a cache by the otherwise unused Canfranc tunnel between France and Spain. Photo courtesy of Juan José Gomez Cadenas.

The process just described, possible only for massive Majorana neutrinos, is not so creatively called *neutrino-less* double β-decay. It has never been observed. Dozens of experiments have tried or are still trying.

The operation of exchanging particles and antiparticles is called *charge conjugation* and labeled *C*. Like parity, *C* is maximally violated by the weak interactions and respected by gravity, QED and QCD. The combination *CP* is also *almost* respected by the weak interactions, that is why very similar things "happen" on the left and right sides of Figure 105, which are related by a *CP* transformation. In the Standard Model, *CP* may be violated and, since the model is a serious theory—in the sense of Chapter 13—this means that it *must* be violated. And it does, in the very way the model predicts.

The operation of inverting "the arrow of time" is called *time reversal*, *T*. It is violated only by the weak interactions, in precisely the same (compensating) amount as *CP* is, since the combined *CPT* operation—again in the sense of Chapter 13—must be absolutely respected. Heroic efforts to observe a violation of *CPT* have so far always failed miserably, as they "should." The *CPT* theorem stating that this combined symmetry of the fundamental interactions is unassailable is one of the two things that are difficult to explain by "waving hand" methods. The other one, closely related to the *CPT* theorem, is the spin-statistics theorem discussed in the first part of Chapter 11.

32

Dark Matter

We have mentioned dark matter many times, without much ado. The summary of what we know about it is fivefold:

- Dark matter acts as particles (elementary or not) with observed couplings to only one of the four known forces: Gravity. It does not couple to photons, thus the name.
- It is "cold": The particles it consists of are currently moving at much smaller velocities than light.
- It dominates the process of the evolution into galaxies and clusters of the small density excesses (above the average) seen in the CBR.
- The gravitational forces exerted by dark matter are observed at various cosmic scales.
- Dozens of experiments searching for non-gravitational effects of dark matter have so far failed. They include direct laboratory searches for interactions with ordinary matter and astrophysical signatures of dark matter decay or annihilation.

Dark energy has observable effects at the largest cosmological distances, probed by the Hubble diagram, and, to a much larger redshift, by the CBR. Dark matter affects the relative height of the acoustic peaks of the CBR, thus the results reported in Figure 94 and Table 1. But very clear dark matter effects can also be found on much smaller scales.

The first truly convincing argument for dark matter was presented by Fritz Zwicky in 1933 in his study of the Coma Cluster of galaxies "close by" to us, at $\sim 3.4 \times 10^8$ light years ($z \sim 0.023$). *Coma,* as it is amiably called, is shown in Figure 108.

In a stable collection of galaxies bound by the gravitational forces they exert on each other, the ensemble's gravitational potential and twice the total kinetic energy must be equal. Zwicky found that this *virial theorem* was not working for the Coma Cluster. From the luminosity of the cluster and the average luminosity-to mass ratio of reference stars in the solar neighborhood, he could deduce the cluster's ordinary-mass

Enjoy Our Universe: You Have No Other Choice. Alvaro De Rújula.

DOI: 10.1093/oso/ 9780198817802.001.0001

Figure 108 The thousands of galaxies of the Coma Cluster. The cluster's angular size is about four times the Moon's. Image credit: Coma Cluster wide-field (ground-based image) heic0813f. Credit: A. Fujii. ©ESA/Hubble.

distribution. But the motions of the galaxies were too fast: They called for a stronger potential and consequently an extra mass that was not luminous.[100] The current estimate is that the Coma cluster's dark mass is larger than its visible one by a factor of ~10.

In 1975, Vera Rubin and Kent Ford announced their discovery concerning the *rotation curves* of galaxies, evidence for dark matter at the much smaller scale of individual galaxies. The rotation curve of the M33 galaxy is shown in Figure 109. As Zwicky did, but much more precisely, one can deduce a mass distribution of stars and gas from the observed light. From this and Newton's law, one predicts the rotational velocity of anything gravitationally bound in this system. The velocities are then

[100] In those bygone days, Zwicky also had to consider that his hypotheses may be wrong; e.g., that the laws of physics may not be universal or that the cluster is not equilibrated.

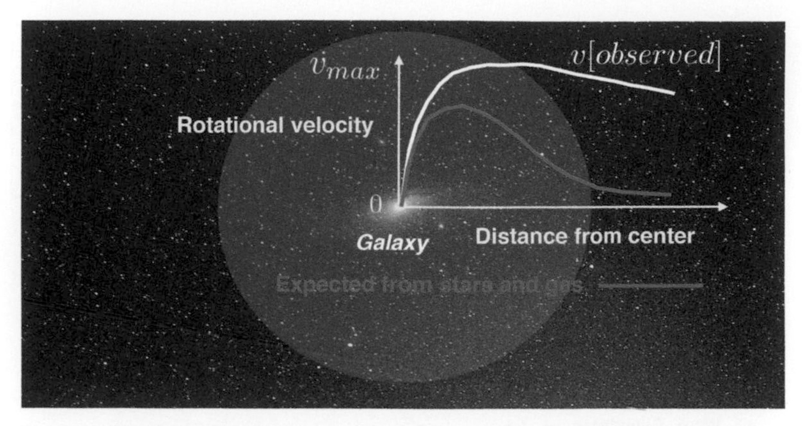

Figure 109 M33 rotation curve. The red curve is the velocity as a function of radial distance, predicted by the mass distribution of ordinary matter, decreasing at (large) R as $1/\sqrt{R}$, like the velocities of planets around the Sun. The white line corresponds to the observed velocities in "21 cm" light, emitted by hydrogen when the spin of its electron and proton flip relative to each other. Illustration by stefania.deluca, Public Domain. https://commons.wikimedia.org/wiki/File:M33 rotation curve HI.gif.

measured: Of stars closer to the galactic center and of small amounts of hydrogen orbiting further away. The prediction fails. Once again, a one order of magnitude extra non-luminous mass is required. Galaxies have dark-matter *halos,* more massive and extensive than their visible part.

Gravitational lensing

A total eclipse of the Sun was visible in May 1919. Expeditions were sent to Brazil and Africa to photograph the stars close to the direction of the Sun when our star was hidden behind the Moon. The results shook the world, read the report by the *New York Times* reproduced on the left column of Figure 110. Why were "men" of science "more or less agog"?

Newtonian mechanics do not apply to objects traveling at the speed of light, like light. Before the advent of relativity and via a somewhat unorthodox Newtonian argument, it was claimed that a body of mass M would bend a ray of light passing at a distance r from its center by an angle $\alpha = 2GM/(rc^2)$, with G being Newton's constant. The correct answer, following from Einstein's equations in Figure 1, is twice that much. The eclipse expeditions vindicated Einstein and gave him

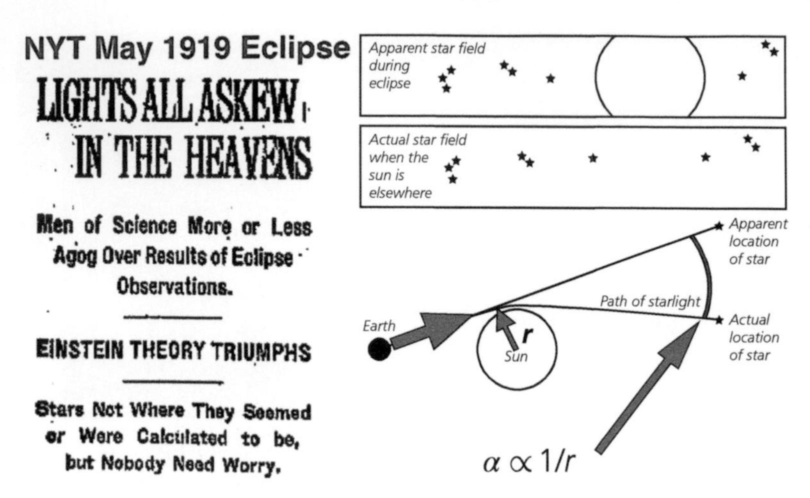

Figure 110 The Sun's gravity modifies the apparent positions of stars in the sky when they are observed (during a solar eclipse, so that the observer is not blinded) in directions close to that of the Sun. The more so, the closer to the Sun they are. NYT headlines: https://en.wikipedia.org/wiki/Predictive power. RHS credit: https://plus.maths.org/content/light-weight, ©NASA.

instant fame. As the *New York Times* then concluded: "... but nobody need worry."

In the eclipse example, the Sun acts as a convergent *gravitational lens.* An extreme example may take place when observing an object that lies behind the lens, precisely in the same direction as its center. The lensed object is then seen completely deformed into what is called an *Einstein ring*, perhaps in honor of Orest Chwolson, who invented the concept twelve years before Einstein. Once again underestimating the unstoppable progress of science, Einstein wrote: "Of course, there is no hope of observing this phenomenon directly. First, we shall scarcely ever approach closely enough to such a central line. Second, the angle α will defy the resolving power of our instruments."

Two of many examples of gravitational lenses giving rise to Einstein rings are shown in Figure 111. When analyzed to extract the mass of the lens, be it a galaxy or a cluster, the result is always that there is an order of magnitude more dark mass than mass in ordinary matter.

Perhaps the most spectacular observation of gravitational lensing by dark matter is the *Bullet Cluster,* shown in Figure 111. Some 150 million years ago two clusters of galaxies had a close encounter.

Figure 111 Left: A nearly perfect Einstein ring of a distant galaxy lensed by a much closer one, built with data by B. Saxton NRAO/AUI/NSF ALMA (NRAO/ESO/NAOJ) Right: *Smilie,* an HST (NASA/ESA) view of a cluster gravitationally lensing more distant (blue) galaxies, ©NASA/ESA. The "eyes" are two bright galaxies.

Figure 112 Bullet Cluster image. Credit: X-ray: NASA/CXC/M.Markevitch et al. Optical: NASA/STScI; Magellan/U.Arizona/D.Clowe et al. Lensing Map: NASA/STScI; ESO WFI.

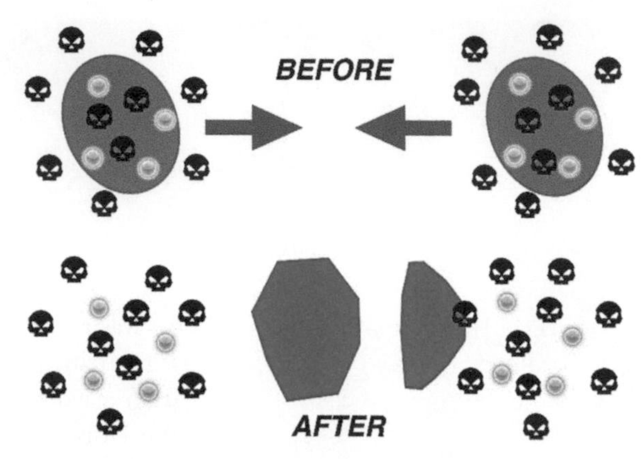

Figure 113 Scheme of the collision of the two clusters of *Bullet*. The gas, shocked after the collision, is orange-colored. Galaxies are represented by the yellow patches. The other symbol stands for dark matter particles.

The ordinary-matter gas in one of the clusters collided with the one in the other, slowing down in their relative motion and now appearing as the red splashes in the figure. The clusters' galaxies and their stars are not few, but they are far between. They did not suffer close encounters and they simply survived, practically undamaged, moving as they did before the clusters met. They are seen now as the galaxies within the bluish patches in Figure 112. These patches, added to the photo a posteriori, are the dark-mass contours deduced from their gravitational lensing effect on background galaxies. The dark mass of the clusters, interacting only gravitationally, also survived the encounter nearly unscathed.

A schematic "before and after" explanation of the interpretation of the Bullet Cluster collision is given in Figure 113.

33

The Origin of "Structures"

Close to the time of recombination, the Universe was an almost uniform soup of ordinary-matter plasma, photons, neutrinos, and dark matter. The former substance became atoms, with a fixed size, subsequently "decoupled" from the expansion of the Universe at larger scales. Under the influence of gravity, the slightly denser regions of dark and ordinary matter gradually became *structures* of stabilized sizes: Stars, planets, galaxies, and clusters. Large clusters are becoming decoupled "as we speak," but much more slowly.

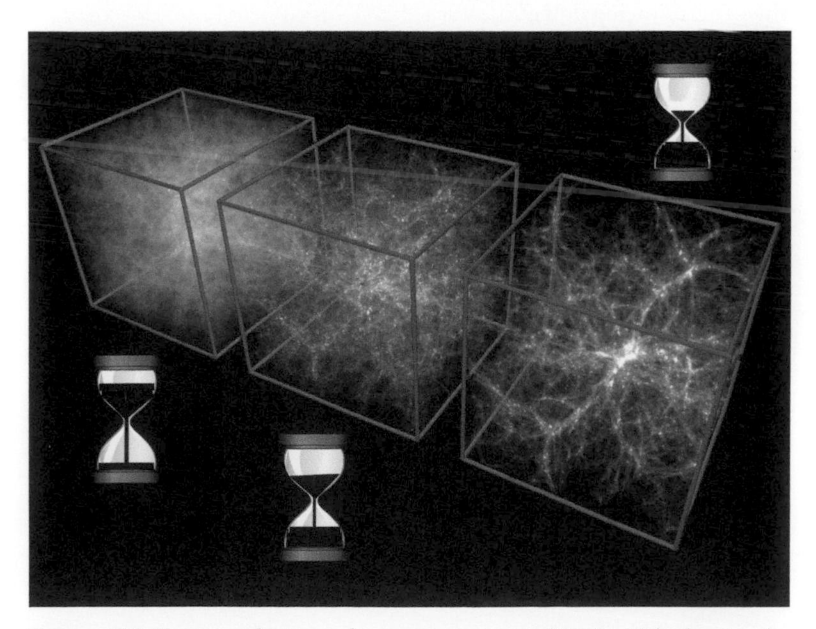

Figure 114 Time evolution of structures in a ΛCDM N-body simulation. Credit: Volker Springel, ©Max-Planck-Gesellschaft zur Förderung der Wissenschaften e.V., Munich.

Enjoy Our Universe: You Have No Other Choice. Alvaro De Rújula.

DOI: 10.1093/oso/9780198817802.001.0001

The study of the formation of structures is relatively simple in a first epoch after recombination. Regions with a density of dark matter slightly bigger or smaller than average evolve under the competing influences of universal expansion and gravitational interactions between dark-matter particles. This increases the density contrast between the more or less dense domains and generates a filamentary structure of denser regions interspaced by "voids." The ordinary-matter contrast simply "follows" the dark matter one. On large scales, the Universe's density contrasts evolve as in Figure 114, which could be an observation, but is actually a computer-generated result.

The evolution pictured in Figure 114 is an example of "*N*-body" simulation[101] in a ΛCDM model, with *N* the number of dark matter particles whose fate the computer tracks, which can reach billions. A computer is necessary as soon as "the regime becomes non-linear," with over- to under-density contrasts significantly larger than unity. Moreover, structure formation at galactic or smaller scales is very complex: Ordinary matter plays a role and its turbulence, cooling or heating by radiation, possible re-ionization, sensitivity to magnetic fields . . . are extremely hard to handle. These issues are still a subject of very active research.

[101] Scientists doing a *computer simulation* do not generally agree with the *Oxford English Dictionary's* definition *1a* of that word: *The action of simulating, with intent to deceive. False pretense, deceitful profession.*

34

The Fate of the Universe

The energy-density of the Universe, as we have seen, is currently dominated by its vacuum energy, recall Figure 94. This energy-density is compatible within observational errors with being isotropic (the same in all directions) and constant in time. That would be Einstein's cosmological *constant*, Λ, something appearing in his general relativistic equations ... somehow. How? Perhaps because in fundamental physics whatever is not forbidden is compulsory. If Λ dominates the Universe now, it will do it forever. That means that the expansion of space will continue to accelerate. Eventually, most clusters of galaxies will be so far from each other that future astronomers will have no way of seeing them. Cosmology will stop being a science and become a chapter of very ancient history.

The primordial inflation required by our current understanding of the Universe may not be *the whole truth* but it is certainly part of the truth, like so many previous chapters in the history of science. The primordial inflation had an end, or the Universe would not have evolved to its current stage. The vacuum energy density responsible for the primordial inflation, thus, was time dependent, it eventually petered out, as discussed in connection with Figure 102. If the current dark energy density is not a constant it may also tend to zero. In that case, the Universe may still expand forever or re-collapse into a Big Crunch. We shall not see any of this, the relevant time scales dwarf the current age of the Universe.

Enjoy Our Universe: You Have No Other Choice. Alvaro De Rújula.

DOI: 10.1093/oso/9780198817802.001.0001

35

Back to the Ether?

To answer a question that is very often posed, let me gather a few remarks about the vacuum, scattered through the book.

The vacuum is a very active substance. Now and in the distant past its energy density controls the expansion of the Universe. At present, a universally uniform Higgs field interacts with all (other) massive elementary particles to endow them with mass. That something remains in an otherwise "empty" space is an idea reminiscent of the ether, the old hypothetical "essence" of space.

The ether was supposed to be the medium in which anything (including light) propagates in an otherwise empty domain, and to be the absolute rest system of space. In this sense, the non-zero vacuum expectation value (VEV) of the Higgs field *is not* an ether. It is a consequence of relativity that *it is not* possible to detect the motion of an object relative to the VEV of a *spin-zero* field. Case closed?

Not quite... it would seem. We know that, carried by the Galaxy, we are moving relative to the system in which the Microwave Background Radiation is maximally isotropic, recall the top panel of Figure 91. Thus there is, after all, a privileged universal rest system in which one can measure one's motion. But this would not be an "absolute" motion, it is relative to the system in which the MWBR is isotropic. It might as well have been relative to a brick. The theory of relativity is doing well and the ether did not strike back.

Enjoy Our Universe: You Have No Other Choice. Alvaro De Rújula.

DOI: 10.1093/oso/9780198817802.001.0001

36

The Crash

> O God, I could be bounded in a nutshell and count myself a king of infinite space, were it not that I have bad dreams.
>
> HAMLET[102]

Studying the micro- and the macro-cosmos—the physical reality at the shortest or largest distances or times—we encounter the same basic concepts: Quantum fields and their interactions. As in Figure 115, we are

Figure 115 The unsuccessful fusion of particle physics and cosmology.

[102] There are hundreds of imaginative interpretations of Shakespeare's work. In an intriguing one, Peter Usher contends that Hamlet is an allegory of the competition in the Bard's times between geo- and heliocentric theories, and of the hybrid one of the (allegedly rotten) Danish astronomer Ticho Brahe. Google "Hamlet cosmic allegory."

Enjoy Our Universe: You Have No Other Choice. Alvaro De Rújula.

DOI: 10.1093/oso/9780198817802.001.0001

Figure 116 A no-longer acceptable attitude toward the cosmological constant: Λ (or something quite akin to it) is measured to be non-zero.

somewhere in the middle of the scale of sizes, with biology, chemistry, and particle physics on one side, geology, astrophysics, and cosmology on the other. But, to date, the merger of the extremes is painful.

A cosmological constant is an energy per unit volume. In the "natural units" of Chapters 3 and 9 that is a mass to the fourth power (M^4). In the equations of Figure 1 describing gravity, the only other constant with mass dimensions is Newton's constant G, with dimensions M^{-2}. This allows one to define a "Planck mass," $M_P = 1/\sqrt{G} \approx 1.22 \times 10^{19}$ GeV, a huge number.[103] If gravity was the only game in town, as during inflation, one would expect the value of the primordial cosmological constant Λ_{inf} to be equal to M_P^4, perhaps times some elegant number, such as 4π. The mass scales of explicit inflationary models do not deviate much from M_P.

Physicists call "unnatural" any number much larger or smaller than unity that one has to introduce to accommodate an observation, for no good theoretical reason. "Decent" theories involve "decent" numbers, such as 8π in Figure 1, 2/3, and −1/3 in Figure 30, 2π in the magnetic anomaly of the electron discussed in Chapter 13, and so on.

The guess $\Lambda \sim M_P^4$ (or thereabouts) would appear to be the "natural" value in any—so far hypothetical—quantum theory of gravity. There

[103] Compare M_P with a proton mass $m_p \approx 0.936$ GeV, or with the current energy of proton-proton collisions at the LHC, 1.3×10^4 GeV.

Figure 117 Top: Photo by Jimmy Harris. Uyuni Train Cemetery, Bolivia. thephysicsmill.com/2015/11/28/. Bottom: A CERN mug in an unappetizing setting, symmetrymagazine.org/07/28/16.

is no known (theoretical) reason why this should not be the expected value for the Universe *right now*. But there is a problem. The value M_P^4 is 120 orders of magnitude larger than the measured current value of the cosmological constant. This is arguably the greatest and (thus) *the most interesting quantifiable conundrum of all time!*

Not so many years ago, the cosmological constant had not been measured and $\Lambda = 0$ was thought to be a possibility.[104] An undiscovered symmetry or mechanism, shown in Figure 116 had to imply this result. Now that Λ has been measured not to be zero the situation is even more challenging. Every physicist should spend a few hours a day trying to explain the observed value of Λ, or be sent to jail.

By now the equations of General Relativity or the ones of the Standard Model of particle physics are so well known that one finds them anywhere, see Figure 117. Yet, nobody has been able to unify them in a grander Theory of Everything. All we have is the acronym, TOE.

[104] The exception was Allan Sandage, for many years Hubble's successor, who insisted in extracting a non-vanishing Λ from his observations.

37

In Spite of Our Admitted Ignorance

Questions come up: Are the concepts such as space and time "basic," or do they "emerge" from something even more "fundamental"? Is our universe a mere "hologram" of something else with a different number of space dimensions? How did inflation start? From a quantum fluctuation, out of nothing at all? How were the laws of Nature born? Are there multiple disconnected universes? If so, do we happen to reside in one of the ones in which intelligent life is possible, or even very probable? There are no known testable scientific answers to such controversial issues, only religious or philosophical beliefs, two of which are reproduced in Figure 118.

As Einstein put it: "We do not only want to understand how Nature works, but we are also after the perhaps utopian and presumptuous goal of understanding why Nature is the way it is and not otherwise."

A most profound mystery of Nature is that it can be pictured and tamed in mathematical terms; the deepest quality of scientists is that they can imagine conceptual abstractions that turn out to be measurable physical realities. The standard models of particle physics and cosmology are examples, and the visionary confidence of some of their developers always makes me recall the very same old story:

My fabulous math teacher in high school was totally embedded in math, in the sense that he fully believed in the actual reality of mathematical abstractions. He would go to the blackboard and draw an x- and a y-axis. From the origin of coordinates his pinched fingers would slowly deploy a fictional z-axis, pointing to the middle of his mesmerized audience. And for the rest of his lecture, as he paced up and down, he would never forget to hold an imaginary line with a steady hand, as he bent to cross under the z-axis, see Figure 119.

Enjoy Our Universe: You Have No Other Choice. Alvaro De Rújula.

DOI: 10.1093/oso/9780198817802.001.0001

The Bible

In the beginning, God created the heavens and the Earth.
The Earth was formless and void, and darkness was over the surface of the deep, and the Spirit of God was moving over the surface of the waters.
Then God said, "Let there be light"; and there was light. God saw that the light was good; and God separated the light from the darkness.
God called the light day, and the darkness He called night. And there was evening and there was morning, one day.

Inflation and before

Basileides of Alexandria, ca. 137CE

Once upon a time there was nothing, nor was there any kind of entity, but in plain language... there was NOTHING AT ALL...

... When there was nothing, neither matter, nor substance, nor non-entity ... no man, nor angel ... nor god... then not-being-god without conciousness or perception, plan, purpose, affection or desire, willed to make a world (I say `willed' to express myself, but I mean an act involountary, irrational and unconcious).

By the 'world' I do not mean the world of time and space which came into being afterwards, but the germ of a world. And this seed contained all things within itself... potentially.

Thus, a not-being god made a not-being world out of nothing.

Figure 118 'The beginning of the Book of Genesis' and an inflationary-like opinion attributed to Basilides (or Basileides). G. Quispel (1968). *Gnostic man: the doctrine of Basilides.* In *The Mystic Vision, Papers from the Eranos Yearbooks,* vol. 6. Routledge & Kegan Paul, London.

Figure 119 Richard Feynman delivering a lecture at CERN on December 17, 1965. In this montage he is impersonating my fabulous math teacher in high school and his axes. The other character standing up is Victor Weisskopf, CERN Director General at the time.

A Concatenated* Glossary of Terms

ANGULAR MOMENTUM A measure of "how much" an object rotates around a certain point. For an object rotating around itself, like a top, the angular momentum is called *spin*.

ANISOTROPIES The slightly hotter or cooler regions observed in the various directions from which the *CBR* reaches us. Evolving after recombination into the current astronomical structures (stars, *galaxies, clusters*) and allegedly originating from quantum fluctuations during *inflation*.

ANTIMATTER/ANTIPARTICLE As a consequence of *Special Relativity* and *Quantum Mechanics*, for every charged particle there exists and antiparticle with the same mass and opposite *charge*. The *electron* and (the predicted) *positron* were the first discovered instance of such a pair.

AXION A very light neutral *boson* predicted by the *Standard Model* and stubbornly escaping detection.

BARYON A non-*elementary particle* made of three *quarks*. Example: The *proton*, $p = (uud)$.

BIG BANG Refers to the origin and evolution of the Universe, which need not be "big" in any conventional sense and is not to be understood as an explosive "bang", but as an expansion of space.

BLACK HOLE An object whose mass-to-radius ratio M/R is large enough for its gravity to impede anything (including light) from escaping from within its "event horizon": A surface at given distance R_s from its center. For a non-rotating hole $M/R > 1/(2G)$, and $R_s = 2GM$, with G Newton's constant, all in *natural units*.

BOSON A particle with integer spin (0, 1, ...). Unlike *fermions*, identical bosons do not "fill boxes". If in the same quantum state (e.g., the same energy and direction of *spin* and motion) adding extra ones is increasingly easier. Lasers work on this principle.

C, THE SPEED OF LIGHT. Roughly one foot per nanosecond (it would be exactly so, for a sensible definition of a foot's length).

CERN The "European Organization for Nuclear Research" near Geneva. A multi-national particle-physics *High Energy Laboratory* not primarily doing "nuclear" physics. Its goals are pure science, its results are all public, non-military, and not pursuant of an "Atomic-Energy" program. Some 2300 "cernois" and 12,000 visiting scientists from over 70 countries and with 120 different nationalities work at CERN. Its cafeterias dwarf the Tower of Babel, though everybody there speaks English. Sort of.

CHARGE See *Electrical Charge* and *Color*.

*Words in italics refer to items defined elsewhere in the list.

CLASSICAL The description of the facts of Nature not requiring (to a given accuracy) the use of use of *Quantum Mechanics.*

CLUSTER A gravitationally bound system of a large number of *galaxies.*

COLD DARK MATTER, CDM *Dark matter* consisting of particles whose *temperature* is negligible, relative to their mass, at the current time in the evolution of the Universe.

COLOR The *charge* of *quarks* and *gluons* that allows them to *couple* to the carriers of the strong "chromo-dynamic" (*QCD*) force, the *gluons*.

COORDINATES The measurements of length in a particularly oriented *reference system*. In Relativity, as is well known, time is treated as a fourth coordinate—or "dimension"—not quite like the others.

CONFINEMENT Of particles with *color: Quarks* and *gluons*. The fact that colored charges cannot be isolated. Particles with *color* occur in Nature only in combinations such that the total color vanishes. The observed confinement is not yet theoretically demonstrated in a fully (mathematically) satisfactory manner.

COSMIC BACKGROUND RADIATION (CBR OR CMB) Light originating in the early Universe and currently permeating it. At present, it has a temperature of ~2.7 K, in the microwave (MW) range of energies. Thus, yet another acronym: *MWBR*.

COSMOLOGICAL CONSTANT Symbolized as Λ. A term in Einstein's equations of *General Relativity*. Interpreted nowadays as the energy density of the vacuum. Allegedly responsible for the current accelerated expansion of the Universe (if Λ is not "it," it is something very very similar).

COSMOLOGICAL PRINCIPLE The hypothesis, confirmed by observations, that the laws of Nature are the same everywhere in the Universe.

COUPLING The interplay between various *elementary particles*, describing their interaction. An $ee\gamma$ coupling, for instance, describes how an *electron* (or its *antiparticle*, the positron) can emit (or absorb) a *photon*; or, in related processes, a photon can become an electron–positron pair, or vice versa.

CRITICAL DENSITY The precise energy density of a universe that, in *General Relativity*, corresponds to its geometry being "flat" (unlike, e.g., the surface of a sphere). A hypothetical expanding universe containing only *matter* is critical if the matter density is finely tuned to gravitationally stop—but not reverse—the expansion.

DARK MATTER An unknown substance made of matter (i.e., particles) and deprived of a significant *coupling* with *photons*, thus not emitting or absorbing light. Only the indirect, collective, astrophysical, or cosmological gravitational effects of dark matter have been observed.

ELECTRIC CHARGE A property allowing an object to *couple* to *photons*, emitting or absorbing them, and being the source of the object's electromagnetic *field*. By convention, the *electron's* charge is negative.

ELECTRON The lightest charged *lepton.*

ELEMENTARY PARTICLE One whose constituents (if it has them) have not yet been discovered. The *Standard Model* description of how they interact with each other is "local": It occurs at a point in space-time. In this sense, elementary particles have vanishing dimensions.

ETHER A hypothetical substance defining an absolute reference system of space and time. Its existence is observationally excluded.

EVENT In relativity, something happening at a specified place and time. In particle physics, the decay of a particle into others, or the results of a collision of two particles.

FERMION A particle with half-integer spin (1/2, 3/2, ...). Unlike *bosons*, identical fermions in the same quantum state (e.g., electrons with the same orbital energy and *spin* direction in an atom) do "fill the box," extra ones are not let in. Thus liquids and solids are very incompressible.

FIELD Here, a "Relativistic Quantum Field." An object "locally" defined at every point of space and time, describing the properties of elementary particles and their interactions. Examples: The electromagnetic field describes electrical and magnetic forces, light waves, and individual photons; the electron field describes electrons and positrons. When *coupled*, the electromagnetic and electron fields describe their interaction.

FIELD THEORY A mathematical description of reality based on the *field* concept. *General Relativity*, *QED*, and *QCD* are examples.

FLAVOR See *Quark.*

FREQUENCY Denoted by ν. The number of times per unit time interval (e.g., a second) that a passing wave is at its maximum. For a wave of velocity v and *wavelength* λ, $\nu = v/\lambda$.

g The acceleration of gravity at the surface of the Earth, $g = G\,M_\otimes / R_\otimes^2 \sim 9.8\,\mathrm{m/s^2}$, with $M_\otimes$ and $R_\otimes$ the Earth's mass and radius.

G The symbol for *Newton's Constant.*

GALAXY A gravitationally bound ensemble of a large number of stars. Our galaxy ("The Galaxy," or the Milky Way) contains a few hundred billion stars.

GAUGE THEORY A *field theory* in which, surprisingly, some basic concepts are partially redundant or unobservable. Example: The voltage of *one pole* of a battery is unobservable, the voltage *difference* between its two poles is. The redundancy: One can add any constant to the voltages of the poles, it disappears from their difference. Though a bit subtler, *QED* and *QCD* are gauge theories.

GENERAL RELATIVITY Einstein's *field* theory of gravity and the geometry of space-time. Based on the local equivalence of acceleration and gravity. Most recent successes: The discovery of the emission of *gravitational waves* by the "Binary

Pulsar" and the detection by LIGO of the ones emitted in a merger of two *black holes*.

GLUON The *spin* 1 carrier of the chromo-dynamic (*QCD*) interactions. Analogous to *photons* but for the fact that the photons are electrically neutral: A photon does not directly *couple* to other photons. Gluons are *colored* and couple to gluons.

GRAVITATIONAL WAVES Ripples in the texture of space-time, traveling at the speed of light, emitted by massive accelerating objects, predicted by *General Relativity*.

GRAVITON The electrically neutral *spin*-two elementary particle mediating gravity. Graviton *couplings* are too weak for individual gravitons to have been observed. From their collective effects we do know their spin and (zero) mass.

HADRON A strongly interacting particle. Either a *baryon* (made of three *quarks*), an antibaryon (made of three antiquarks), or a *meson* (made of a quark and an antiquark).

HEAVY LEPTON A clear oxymoron: *Lepton* stands for a (relatively) light *elementary particle*. The τ is a heavy lepton.

HIGH ENERGY LABORATORY Refers to experimental facilities studying particle physics, mainly at high energies. The largest ones at the moment are *CERN*, straddling Switzerland and France, and Fermilab in the United States. Before and after World War II, particle physics—and so many other intellectual activities in Europe—had to move to the USA, where several labs were developed, such as BNL (at Brookhaven, New York State) and SLAC (at Stanford, California). With the tragic demise of the SSC (an accelerator project larger than the LHC, at Waxahachie, Texas) much of this experimental physics activity moved back across the Pond.

HIGGS BOSON A neutral *spin-zero* particle (almost) completing the *Standard Model of elementary particles*. Almost? See *Axion*.

HIGGS FIELD The *Relativistic Quantum Field* describing the *Higgs boson* and participating in the description of its interactions. Its value in the *vacuum* is not zero (the vacuum is full of it). Interacting with this (non)vacuum, *elementary particles* acquire their mass. The non-zero vacuum value of this (*spin* 0) field does not constitute an *ether*. The naively estimated energy density of the Higgs field is many orders of magnitude bigger than the measured *cosmological constant:* A serious conundrum.

INERTIAL Not subject to acceleration. Referring to an observer or the spacial *reference system* in which (s)he is at rest. This may include spacial *coordinates* and clocks.

INFLATION An exponentially accelerated expansion of the space of the Universe. "Exponential" in the sense that the "scale" of space (e.g., the distance between two faraway galaxies) repeatedly doubles after successive fixed intervals of time. The Universe is observed to be now inflating... again (see next entry).

INFLATIONARY PARADIGM The observationally supported view that the Universe underwent a period of *inflation* in the first tiny fractions of a second of its evolution.

INTERMEDIATE VECTOR BOSONS The *spin*-one carriers of the weak interaction or force. The Z is electrically neutral. The W^+ and W^- are a particle/antiparticle pair.

ION An atom deprived of one or more of its electrons.

KINETIC ENERGY The energy of motion. For a particle of mass m and velocity v much smaller than c, $E_k \approx (1/2)\, mv^2$.

ΛCDM The current Standard Model of cosmology. See *Cosmological Constant* and *Cold Dark Matter*.

LEPTON A *spin* 1/2 elementary particle with no *color*. The electron, e; muon, μ; and "tau," τ, are electrically charged leptons. "Their" corresponding *neutrinos* are not (see *Neutrino*).

LORENTZ FACTOR The function $\gamma(v) = 1/\sqrt{1 - v^2/c^2}$, an ubiquitous factor in formulae of *Special Relativity*.

MATTER In cosmology, any non-*relativistic* constituent of the Universe.

MESON A non-*elementary particle* made of a *quark* and an antiquark; for example, $\pi^- = (d\bar{u})$. The etymology (neither light nor heavy) is obsolete.

MICROWAVE BACKGROUND RADIATION (MWBR) The *Cosmic Background Radiation* at its current *temperature*.

NATURAL UNITS An extremely convenient concept. Choose c, the velocity (distance over time) of light to be $c = 1$. Then time (seconds) and distance (seconds... of light travel) have the same units. And so do energy and mass (recall $E = mc^2$). Adding the choice of a reduced Planck constant, $\hbar = 1$, energies and frequencies have the same units, *spin* is specified by numbers with no units. All quantities are expressible in powers of mass. Add $G = 1$ and all units disappear (but for money).

NEUTRAL CURRENT A *weak interaction* mediated by the exchange of a Z, the neutral *Intermediate Vector Boson*. For example, $\nu_e\, p \to \nu_e\, p$. See the next entry.

NEUTRINO An electrically neutral *lepton*. In a "charged current" weak interaction an "electron neutrino," ν_e, becomes an electron (or a $\bar{\nu}_e$ a positron). Similarly a ν_μ becomes a muon and a ν_τ a τ *lepton*.

NEUTRON One of the two constituents of atomic nuclei, itself made of two *down* and one *up quarks*, $n = (udd)$.

NEWTON'S CONSTANT Symbolized as G. Specifies the strength of the gravitational force F between two (non-*relativistic*) objects of masses M_1 and M_2, separated by a distance R. Explicitly: $F = -\, G M_1 M_2 / R^2$. The minus sign denotes attraction.

NUCLEON A *proton* or a *neutron*, the constituents of atomic nuclei.

NUCLEUS The positively charged object at the center of an atom, made of *protons* and **nucleons** and having a radius of a few times 10^{-5} of the atom's radius.

NEUTRON STAR A celestial object of mass comparable to the Sun's, made almost exclusively of neutrons, and as dense as an atomic nucleus. In a sense, these stars are a continuation of the Periodic Table of the elements.

PHOTON The particle of light and of the non-visible electromagnetic radiation. Microwave, radio, and infrared photons have lower-than-visible energy or frequency. Ultraviolet photons, X-rays, and gamma-rays have higher-than-visible energy.

PION A spin zero boson made of a light *quark* (*u* or *d*) and a light antiquark; for example, $\pi^+ = (\bar{d}\,u)$.

PLASMA The fourth state of matter (besides solid, liquid, and gas). Dense and hot enough for its components (atoms or even nuclei and *nucleons*) to be broken into their constituents by mutual collisions.

POSITRON The *antiparticle* of the *electron*, having the same mass and *spin* as the electron, but the opposite charge.

POWERS OF 10 Femto, nano, micro, milli, . . . , kilo, Mega, Giga, Tera, stand for 10^n, with $n = -12, -9, -6, -3, \ldots, 3, 6, 9, 12$.

PROTON One of the two constituents of atomic nuclei, itself made of two *up* and one *down quarks*, $p = (uud)$.

PULSAR A rotating *neutron star* emitting electromagnetic radiation (light, X-rays . . .), beamed as that of a lighthouse.

QUANTUM CHROMODYNAMICS (QCD) The relativistic quantum *field theory* of particles carrying a *color charge*: *Quarks* and *gluons*.

QUANTUM ELECTRODYNAMICS (QED) The relativistic quantum *field theory* of *photons* and electrically *charged* particles.

QUANTUM MECHANICS OR QUANTUM THEORY A stunning but extremely well-tested ingredient in the description of Nature. "Quantum" describes the fact that certain observables—such as the energies of photons emitted by atoms—have values that are not continuous. Also, the product of certain pairs of observables—such as the position and momentum of a particle—cannot be smaller than a specific "quantum" value.

QUARK A spin 1/2 elementary particle occurring in six different *flavors*. "Flavor" is a sophomoric appellation for the quarks of different mass and electric charge. The six quark flavors are *u*, *c*, *t* (of electric *charge* 2/3) and *d*, *s*, *b* (of *charge* −1/3). Each quark flavor occurs in three different *colors* with identical masses. The *s*, *c*, *b*, and *t* quarks are far more massive than *u* and *d*, they are extremely unstable; for example, $s \to u\bar{u}d$.

QUASAR A gigantic *black hole* at the center of a *galaxy* accreting matter, a fraction of which is spat out in the form of jets that emit visible electromagnetic radiation, most intense and energetic in the jets' directions.

RECOMBINATION The process in which, when the age of the Universe was about 380,000 years, ordinary matter ceased to be a *plasma* (mainly made of *protons electrons* and *photons*) to become atoms (mainly hydrogen). At this epoch the Universe became transparent to the photons that we now see as the *MWBR*.

REDSHIFT Denoted by z. A measure of the amount by which the wavelength of light is stretched by the receding motion of the light's source, or by the expansion of the space in which the emitter and observer are immersed. The first possibility is well known in its sound analog.

REFERENCE SYSTEM, OR FRAME An actual or abstract system of *coordinates* and the set of reference points, "milestones," which uniquely locate and orient the *coordinate system*. In relativity, this includes clocks at rest at the milestones.

RELATIVISTIC A theory obeying Einstein's relativity, or an object traveling at a velocity not much smaller than the velocity of light.

SPECIAL RELATIVITY Einstein's theory of space and time and of the motions of light and/or massive objects, moving at constant velocity *relative* to each other. It deals with *inertial* observers and their *reference systems*.

SPIN A quantum-mechanical concept describing how an object "behaves" under rotations or, somewhat incorrectly, how it rotates about itself. Even *elementary particles*—with no substructure—may have a non-vanishing spin. Spin is "quantized." In *natural units* it can be an integer (0, 1, ...) or a half-integer (1/2, 3/2, ...). *Bosons* have integer spin, *fermions* a half-integer one.

STANDARD MODEL (OF ELEMENTARY PARTICLES) A theory describing the strong (*QCD*) interactions as well as the unified electromagnetic (*QED*) and *weak interactions*. Latest success: The discovery of the *Higgs boson*.

STRONG INTERACTIONS Currently meaning the chromo-dynamic (*QCD*) forces. In the past the expression referred to the forces between *protons*, *neutrons*, and other *hadrons*, forces no longer considered fundamental.

SUPERNOVA The violent explosion of an aging massive star or of a fusing two-star system. Very luminous supernovae of the latter type are used to measure cosmic distances.

TEMPERATURE The average kinetic energy of the constituents of a substance. Even an ensemble of *photons* (such as the *MWBR*) can have a "thermal distribution" of energies, characterized by a temperature (total and kinetic energy are the same for photons, which are massless).

VACUUM What remains in a space emptied of everything that can be taken away. The vacuum is observed not to be truly empty: Not everything can be taken away from it. The vacuum is a substance, currently permeated by a *Higgs field* of a magnitude constant in space and time.

WAVELENGTH Denoted by λ. The distance between two successive crests of a wave. For a wave of velocity v and *frequency* ν, $\lambda = v/\nu$.

WEAK INTERACTIONS The ones mediated by *Intermediate Vector Bosons:* Z, W^+, and W^-. These *spin-one* force carriers are analogous (and related) to *photons*, but their masses are non-zero.

VECTOR A quantity with both a magnitude and, in a specific *reference system*, a direction in which it points.

Index

Figures and captions are indicated by an italic *f* following the page number. Tables are indicated by an italic *t* following the page number. Footnotes and glossary items are indicated by an italic *n* or *g* following the page number. Main sections for a topic are indicated in bold.